This book aims to introduce the reader to the behaviour of electrons in solids, starting with the simplest possible model, and introducing higher-level models only when the simple model is inadequate.

Unlike other solid state physics texts, this book does not begin with complex crystallography, but instead builds up from the simplest possible model of a free electron in a box. The approach is to introduce the subject through its historical development, and to show how quantum mechanics is necessary for an understanding of the properties of electrons in solids. It does not treat the dynamics of the crystal lattice, but proceeds to examine the consequences of collective behaviour in the phenomena of magnetism and superconductivity. Throughout the mathematics is straightforward and uses standard notation.

This text is suitable for a second- or third-year undergraduate course in physics, and would also be suitable for an introductory solid state course in materials science or materials chemistry.

INTRODUCTION TO THE PHYSICS OF ELECTRONS IN SOLIDS

INTRODUCTION TO THE PHYSICS OF ELECTRONS IN SOLIDS

BRIAN K. TANNER
Professor of Physics
University of Durham

Published by the Press Syndicate of the University of Cambridge
The Pitt Building, Trumpington Street, Cambridge CB2 1RP
40 West 20th Street, New York, NY 10011-4211, USA
10 Stamford Road, Oakleigh, Melbourne 3166, Australia

First published 1995

Printed in Great Britain at the University Press, Cambridge

A catalogue record for this book is available from the British Library

Library of Congress cataloguing in publication data
Tanner, B. K. (Brian Keith)
Introduction to the physics of electrons in solids/Brian K. Tanner.
p. cm.
ISBN 0 521 23941 9 – ISBN 0 521 28358 2 (pbk.)
1. Solid state physics. 2. Energy-band theory of solids.
3. Semiconductors. I. Title.
QC176.T32 1995
530.4'11–dc20 94-11666 CIP

ISBN 0 521 23941 9 hardback
ISBN 0 521 28358 2 paperback

KT

To my mother
In memoriam

Contents

Preface

Most textbooks on solid state physics begin the exposition from what might be called a 'structural position'. Space and point groups are discussed, followed by consideration of the Bravais lattice. The reader is thus led on to elementary ideas about crystallography and the use of diffraction techniques for the solution of crystal structures. Having laid the foundation of how atoms and molecules order to form crystalline structures, electron motion in such periodic structures is treated and band theory developed. The free electron model is seen as an approximation of the more general band theory. In many, rather formal, ways this approach is very satisfying. It would seem obvious that in the first instance one must understand the structure of the material on which one is working before attempting to understand its other physical properties. However, in practice, it proves rather hard to teach solid state physics this way and to retain student enthusiasm in the early stages of the teaching of crystallography where one is dealing with rather difficult geometrical concepts and very little physics. There is a very real danger of making the introduction to the subject so unexciting that the inspiration is lost and students come to regard solid state physics as the 'dull and dirty' branch of their physics course. However, elementary quantum mechanics, including the one-dimensional solution of the time independent Schrödinger equation, is included quite early in many undergraduate courses and there is much attraction in illustrating at this early stage the important technological context of the apparently abstruse quantum mechanics. Further, in solid state physics one is dealing with very many particles, and one is forced to make approximations. Exact solution of the equations of motion of 10^{22} particles is clearly out of the question despite the fact that we understand the electromagnetic interaction so well! Thus in the first solid state physics course at Durham we have tried both to utilize quantum mechanics at as early a stage as possible and also to introduce some concepts of model building. To this end we have taken the results of the solution of Schrödinger's equation for a single particle in a one-dimensional potential well, extended them and applied them to the problem of an electron in a solid. In this way, the student meets first the simplest possible model for the behaviour of electrons in solids, the free electron approximation where the potential well is flat bottomed and infinitely deep. The results are compared with experimental observations, and only when there is clear discrepancy between theory and experiment is the model made more complex. In many respects this represents a historical approach and one which is all too rarely adopted. There is a great danger that undergraduates come to regard their physics as the ultimate description of the universe which somehow appeared abruptly as a complete, well formulated, entity. While the teaching of relativity goes some way to correcting this impression, students rarely appreciate how physics is a constantly evolving science and how one is all the time building models to describe the physical world. Model building is at the heart of modern physics and is not just restricted to elementary particle physics. Accordingly the approach to the teaching of the physics of solids adopted in this book hinges on two principles. Firstly to illustrate something of the historical development of this subject and to show how necessary is quantum mechanics to understanding the behaviour of electrons in solids. Secondly, to begin with the simplest possible model and gradually increase the complexity. Thus the Drude classical free electron theory is followed by quantum mechanical free electron theory which in turn leads into elementary band theory as a perturbing periodic potential is introduced. The first six chapters are concerned with itinerant electrons under the independent particle approximation, developing the free electron model into elementary band models. In order to illustrate the technological application of the results, Chapter 7 is devoted to an examination of the physical properties of several solid state electronic devices. Chapter 8 examines the behaviour of localized electrons while the last two chapters are devoted to phenomena resulting from breakdown of the independent electron approximation, namely ferromagnetism and superconductivity.

The book is aimed at first or second year undergraduate level and a genuine attempt is made to keep the mathematical discussion as simple as is consistent with clarity. It will, it is hoped, have appeal not only to students of physics but also of chemistry, metallurgy, geology and engineering. The text is based on a lecture course given originally to a mixed class of physicists, chemists, mathematicians and geologists at Durham University and the general response to the approach has been favourable. In restricting the scope and length of the text, it is hoped that individual students will be able to use it as a genuine introduction to the physics of solids, before graduating to one of the more weighty and comprehensive 'introductory' texts on the market.

Brian Tanner

Acknowledgements

My sincere thanks are expressed to the many people who have assisted me in the preparation of this book. I am particularly grateful to Pauline Russell who patiently drew the diagrams from my spider-like sketches, to Nikki Bingham who has deciphered my handwriting with patience and care, to Vikki Greener who has uncomplainingly photocopied large quantities of material and Mike Lee whose skills with the camera continue to amaze us. Of my colleagues, Dick Fong has been particularly assiduous in pointing out errors in problems set over the years and I am indebted to the generations of students who have acted as stoical guinea pigs for my ideas, explanations and exercises. Finally, I acknowledge the love and support of my wife Ruth and sons Rob and Tom, who have tolerated my eccentricities and obsessions with equanimity and humour.

1

The classical free electron model

In practical terms, the enormous range of values of resistivity of solids is something which we take for granted. Every day, we happily touch the polymer sleeving or fitments surrounding conductors bearing potentially lethal currents at quite high voltages. Only when one is reminded that the difference of almost 30 orders of magnitude found between the resistivity of the noble metals and some synthetic polymers represents the largest variation of any physical parameter does this apparently mundane phenomenon suddenly appear intriguing. It is tempting to enquire whether the same physical process can be responsible for electrical conduction in all materials. Even if the same process extends over half the range it would be a remarkable achievement. Perhaps we would then appreciate why so much time and effort is devoted to measurements of electrical conductivity.

It is hard to conceive that, prior to the turn of the century, very little was known about the physics of solids. Some ideas on crystal structure had been anticipated from the morphology of natural and synthetic crystals but there existed little understanding of the electrical, thermal or magnetic properties of solids. Solid state physics is a twentieth century branch of science and as such deserves recognition as an important section of 'modern physics'. As we shall see later, it was not until quantum mechanics was applied to the physics of solids that many real advances were made.

The first major step in our understanding of the electrical properties of solids was taken by Drude in 1905. Around this time, J. J. Thompson had been performing his classic experiments on the properties of cathode rays. These cathode rays, which were electrically negative, were produced by heating a metallic filament. Somehow, the cathode rays were 'boiled off' from the metal. What was most important was that the properties of the cathode rays were independent of the metal used for the filament. They seemed to be contained in all metallic solids. Drude's suggestion was that these cathode rays, which we now know as electrons, might be responsible for the electrical conduction in metals. As it turned out, this was a very perceptive insight.

The theory which Drude developed was, naturally, based on classical principles. In it is made an important assumption which is implicit throughout the first eight chapters of this book. This is the INDEPENDENT ELECTRON APPROXIMATION. Drude himself made the approximation, but it is so important that it deserves highlighting here before any further discussion takes place. From what we know about the spacing of atoms in solids, following the pioneering X-ray diffraction experiments of Friedrich and Knipping in 1912, it becomes obvious that a 1 cm cube of metal contains a very large number of electrons. In principle we should consider the interaction of each electron with all the others in the solid via the Coulomb electrostatic interaction. This is clearly impossible to do in practice, so what is done is to assume that each electron moves in an average potential created by all the other electrons, as well as the positive ions in the crystal. Thus we can treat the motion of one electron independently and simply add the contributions of the individual electrons to get the collective response. We can handle the physics of one particle in an average potential well but we cannot handle many particles simultaneously. The independent electron approximation underlies the whole treatment of electronic behaviour developed in the first eight chapters. When it breaks down, as discussed in the last two chapters, very important phenomena arise.

In the classical free electron theories, including that formulated by Drude, one makes the following assumptions in addition to the independent electron approximation.

1 Conduction is entirely by electrons.
2 The sample defines a flat bottomed potential well within which the electrons are constrained and within which they are free to move.
3 The electrons behave as a classical gas, i.e.
   (a) they are distinguishable
   (b) they are small and take up negligible volume
   (c) they have random motion
   (d) they are perfectly elastic.

4 There are no quantum restrictions on the electron energy i.e. the energy distribution of the electrons is a perfect continuum.

1.1 Drude theory

In addition to the above assumptions, Drude made the assumption that all electrons have the average energy. As can be seen in Appendix 2, application of statistical mechanics to a large ensemble of classical particles predicts a probability distribution for the electron energy known as the Boltzmann distribution and the Drude assumption is clearly a very drastic approximation. However, the Drude model is the *simplest possible* model for the behaviour of electrons in metals and therefore we examine it first.

1.1.1 Electrical conductivity (σ)

Suppose that an electron, labelled i, mass m^* is acted upon by a field $\mathcal{E}_x$ in the x direction. This results in a force of $e\mathcal{E}_x$ acting on the electron in the x direction, where e is the charge on an electron. From Newton's Second Law, this equals the rate of change of momentum in the x direction. Thus, if v_x is the x component of velocity of the electron, we have

$$m^* \mathrm{d}v_{ix}/\mathrm{d}t = e\mathcal{E}_x. \qquad (1.1)$$

Now the right hand side of this equation is the same for all electrons and hence we can write

$$m^* \mathrm{d}\langle v \rangle/\mathrm{d}t = e\mathcal{E}_x, \qquad (1.2)$$

where $\langle v \rangle$ is known as the drift velocity in the x direction and is given by

$$\langle v \rangle = \frac{1}{n}\sum_{i=1}^{n} v_{ix} \qquad (1.3)$$

and n is the number of electrons considered.

The drift velocity is essentially different from the random velocities of the individual electrons as it represents a net motion in the field direction superimposed upon such random motion. Fig. 1.1 gives an example of such a drift in the x direction. In zero electric field, on average the electron does not change its position despite its random motion. In a non-zero electric field, the position in the field direction changes, and the electron drifts in this direction. Clearly this leads to a

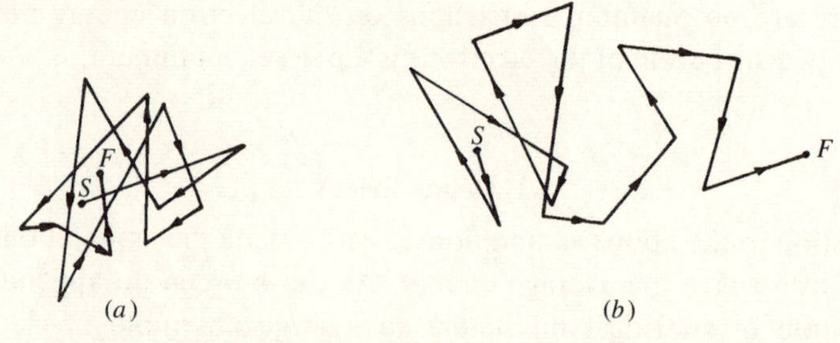

Fig. 1.1 Concept of drift velocity. Schematic electron motion in (*a*) zero (*b*) non-zero electric field. In an electric field, the start, *S*, and finish, *F*, positions differ, leading to a net flow of charge.

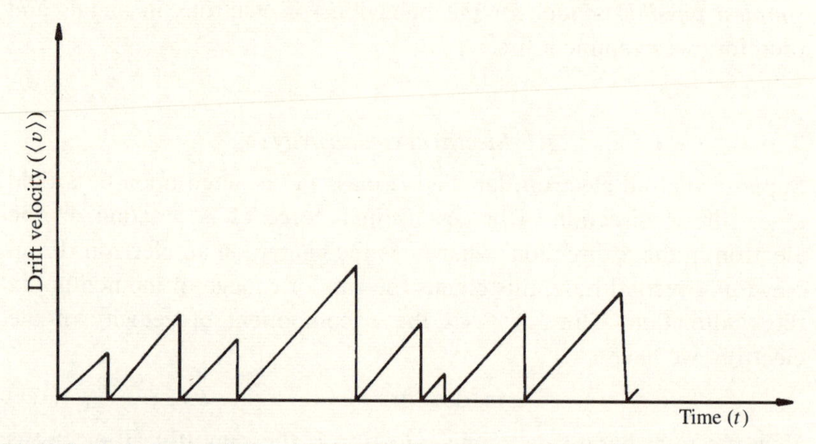

Fig. 1.2 Variation of drift velocity with time. Average time between collisions is τ, the relaxation time.

net flow of charge and hence the passage of an electric current. According to Equation (1.2), the drift velocity $\langle v \rangle$ should increase linearly with time if the electric field is constant. This implies a linear increase of current in a constant field, which is at odds with experimental observation.

The dilemma is resolved by the assumption that the electrons encounter obstacles to their motion and that after such collisions, the motion is randomized and the drift velocity destroyed. After a collision all memory of the previous motion is lost. In effect the clock is set back to zero. The drift velocity will then vary with time in a manner illustrated in Fig. 1.2. On average, collisions occur every time interval of τ, which is known as the relaxation time. As $m^*\langle v \rangle$ of momentum is destroyed at each collision, the rate of destruction of momentum is

$m^*\langle v\rangle/\tau$. There is thus an effective force of this magnitude slowing down the electrons.

We can then write a general equation of motion for the drift motion,

$$m^*\mathrm{d}\langle v\rangle/\mathrm{d}t = e\mathcal{E}_x - m^*\langle v\rangle/\tau. \tag{1.4}$$

The left hand side represents the net rate of change of drift momentum, the first term on the right hand side is the electrostatic force and the second is the 'viscous drag' due to collisions with the obstacles. In the steady state the left hand side is zero. Hence,

$$\langle v\rangle = e\tau\mathcal{E}_x/m^*, \tag{1.5}$$

Now the current density J is defined as

$$J = ne\langle v\rangle, \tag{1.6}$$

where n is the number of electrons per unit volume. Thus

$$J = ne^2\mathcal{E}_x\tau/m^*, \tag{1.7}$$

and as the conductivity σ is defined as

$$\sigma = J/\mathcal{E}_x \tag{1.8}$$

we have

$$\sigma = ne^2\tau/m^*. \tag{1.9}$$

You will notice that Equation (1.7) is a proof of Ohm's Law; that the current is proportional to the electric field (and hence voltage). As expected, σ varies linearly with n, the electron density and also with the relaxation time τ.

So far, so good, but the test of any theory in solid state physics is to check its temperature dependence. According to our model, all the electrons have the average energy and so will be travelling with the same (random) velocity which we call V. Then if the mean free path, the mean distance between collisions, is l we have

$$l = V\tau. \tag{1.10}$$

It is straightforward to show from classical kinetic theory (Appendix 1) that

$$m^*V^2/3 = k_\mathrm{B}T, \tag{1.11}$$

where k_B is Boltzmann's constant and T is the absolute temperature. Thus

$$\sigma = ne^2l/(3m^*k_\mathrm{B}T)^{1/2}. \tag{1.12}$$

This is the first important result of the Drude theory.

1.1.2 Thermal conductivity (K)

While it is a common observation that the best electrical conductors are also the best thermal conductors, the wide variation found in electrical conductivity is not displayed for thermal conductivity. A certain amount of thermal conduction occurs in even the worst electrical conductors and this is associated with lattice vibrations. However, the very high thermal conductivities of metals suggest that the bulk of the thermal conduction is by electron transport.

Let us assume that a temperature gradient $\mathrm{d}T/\mathrm{d}x$ exists along a rod of metal. At temperature T, each electron has an average energy of $3k_\mathrm{B}T/2$. (This follows from the equipartition of energy theorem.)

An electron in a hotter region at temperature $T + \delta T$ has energy $3k_\mathrm{B}T/2 + 3k_\mathrm{B}\delta T/2$. This excess energy tends to make it diffuse to a cooler region and changes as we go along the specimen as if a potential were acting on the electrons. This thermodynamic potential gives rise to an effective thermodynamic force $3/2k_\mathrm{B}\mathrm{d}T/\mathrm{d}x$.

The thermodynamic force gives rise to drift motion which is limited by collisions of the electrons with obstacles. The equation of drift motion now becomes

$$m^*\frac{\mathrm{d}}{\mathrm{d}t}\langle v\rangle = \frac{3}{2}k_\mathrm{B}\frac{\mathrm{d}T}{\mathrm{d}x} - m^*\frac{\langle v\rangle}{\tau}. \tag{1.13}$$

Notice that the relaxation time τ is the same as before. In the steady state

$$\langle v\rangle = \frac{3k_\mathrm{B}\tau}{2m^*}\left(\frac{\mathrm{d}T}{\mathrm{d}x}\right). \tag{1.14}$$

Each electron carries energy of $3k_\mathrm{B}T/2$. Thus the heat current U per unit area is given by

$$U = n\left(\frac{3k_\mathrm{B}T}{2}\right)\left(\frac{3k_\mathrm{B}T}{2m^*}\right)\left(\frac{\mathrm{d}T}{\mathrm{d}x}\right). \tag{1.15}$$

Therefore, the thermal conductivity K, which is defined as

$$K = U/(\mathrm{d}T/\mathrm{d}x), \tag{1.16}$$

is given by

$$K = 9k_\mathrm{B}^2 n\tau T/4m^* \tag{1.17}$$

or, in terms of the mean free path,

$$K = 9k_\mathrm{B}^2 nlT/4m^*V. \tag{1.18}$$

According to classic kinetic theory

$$K = (3\sqrt{3})k_\mathrm{B}^{3/2}nlT^{1/2}/4(m^*)^{1/2}, \tag{1.19}$$

and we thus expect the thermal conductivity to vary as the square root of the absolute temperature.

One important point has been glossed over in this derivation. Unlike the case of electrical conductivity, where the electrons flow out of one end of the sample and are returned to the starting point through the external circuit, in the thermal case we must conserve electrons. The net electron flux across any plane must be zero. (Note that although there is no net mass transport, there can still be a net momentum transport as the electrons travelling from the hotter region towards the cooler region have more energy then those travelling in the opposite direction.) Initially, on setting up the temperature gradient, there will be a net flow of electrons to the cooler end creating a concentration gradient. This in turn acts to increase the number of cooler electrons diffusing towards the hotter region and thus a situation of dynamic equilibrium is reached.

The number of particles F crossing unit area per unit time down a concentration gradient $\mathrm{d}n/\mathrm{d}x$ is given by Fick's Law,

$$F = -D(\mathrm{d}n/\mathrm{d}x), \tag{1.20}$$

where D is the diffusion coefficient. In the steady state we must have zero net flux across the plane, i.e. the fluxes due to temperature and concentration gradients are equal. Thus,

$$n\langle v\rangle = -D(\mathrm{d}n/\mathrm{d}x). \tag{1.21}$$

This is a specific integral form of the more general 'continuity equation'.

One can thus see that with varying electron concentration the problem becomes, in principle rather complex. However, it is common experience that connecting a thin conductor across the ends of a metal specimen does not appreciably alter the thermal properties. We therefore conclude that the concentration gradient is small, and our theory adequate.

1.1.3 Thermoelectric effects

If however, we do consider a complete circuit we observe some rather small electrical effects associated with heat transport known as thermoelectric effects. Suppose two specimens A and B of different metals are connected together as shown in Fig. 1.3. The temperature of the circuit is kept constant at T and an electric current, of current density J is passed round the circuit. Associated with the electric current is

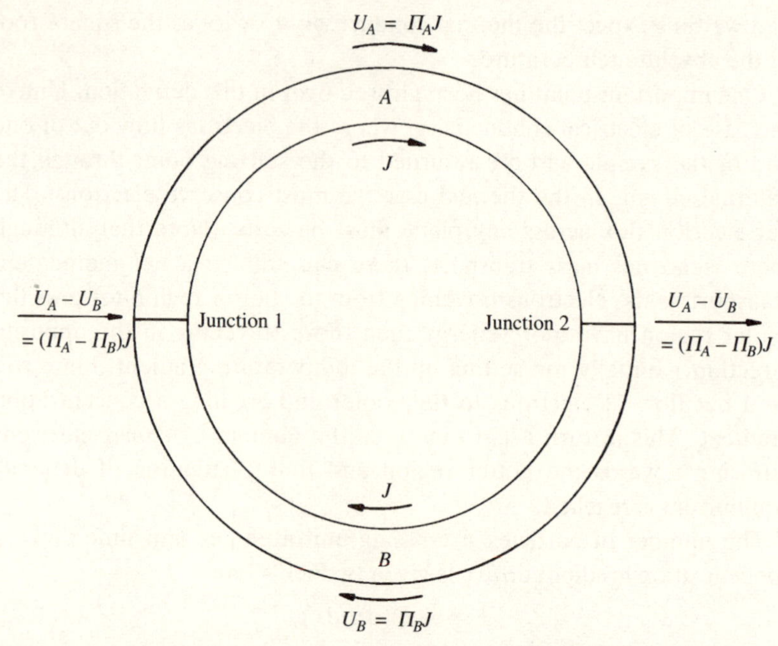

Fig. 1.3 The thermoelectric effect in which heat is absorbed or rejected at a junction in a ring of different metals supporting an electric current.

also a heat current, as each electron carries an average energy $3k_BT/2$. The heat current density U is related to the electrical current density J by the Peltier coefficient Π,

$$U = \Pi J. \tag{1.22}$$

Now if $\Pi_A \neq \Pi_B$, as is usually the case, the heat currents in the two sections A and B will differ and in order to preserve continuity heat must be absorbed and rejected at the junctions. In Fig. 1.3 we have chosen $\Pi_A > \Pi_B$. By measuring the heat given out at such junctions we may measure the Peltier coefficients of metals. From Equations (1.6) and (implicitly) (1.15) we find that,

$$\begin{aligned} \Pi = \frac{U}{J} &= n\frac{(3k_BT/2\langle v\rangle)}{ne\langle v\rangle} \\ &= 3k_BT/2e. \end{aligned} \tag{1.23}$$

1.1.4 Wiedemann–Franz Law

From Equations (1.9) and (1.17) we note that

$$K = \sigma(3k_B/2e)^2T = \sigma LT \tag{1.24}$$

where

$$L = (3k_B/2e)^2. \tag{1.25}$$

L is known as the Lorenz number and is independent of electron concentration.

1.1.5 Electronic specific heat capacity

Each electron carries the average energy of $3k_BT/2$ and hence the total electron energy, E, is $3nk_BT/2$. Thus we expect a contribution to the specific heat capacity at constant volume C_e, which is defined as $(dE/dT)_v$, given by

$$C_e = 3nk_B/2. \tag{1.26}$$

1.2 Experimental tests of the Drude theory

We now have a number of predictions based on the Drude theory which must be examined in the light of the experimental evidence. The first and most important question to be asked concerns the validity of the assumption that the charge carriers are electrons. Several elegant experiments have been performed to test this assumption, of which one will be described here and another features as a problem at the end of the chapter.

1.2.1 Charge to mass ratio of current carriers – Tolman and Stewart experiment

The first definitive experiment to determine the charge to mass ratio of the current carriers in a metal was reported by Tolman and Stewart in *Physical Review* **9** (1917) 164. It is based on a comparison of the inertial energy of rotating electrons with the electrical energy dissipated on stopping them. The inertial energy involves mass; the electrical energy involves charge.

As illustrated in Fig. 1.4, a coil of length Z is wound round the perimeter of a disc of radius r. The disc rotates with constant angular velocity ω_0 and is braked rapidly to a halt. As the electrons are effectively free inside the metal, they continue to rotate after the wire has come to a standstill. A current is thus observed to flow in a ballistic galvanometer attached across the ends of the coil by means of slip rings. Assuming that there are p turns of wire and N is the number of

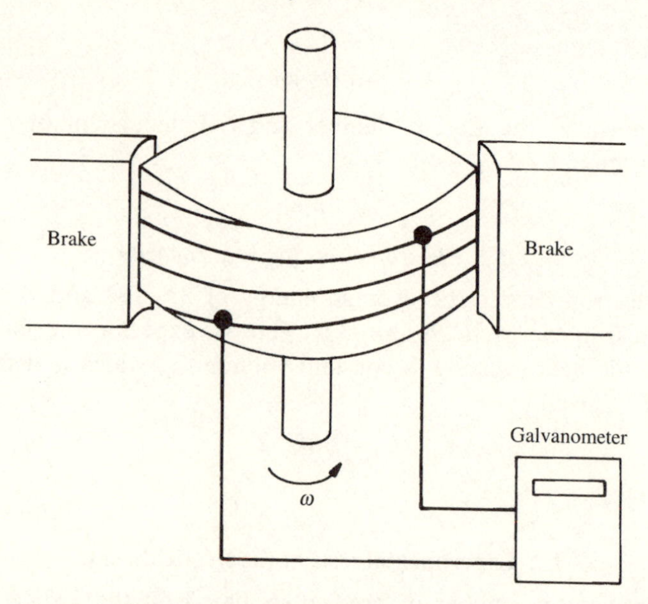

Fig. 1.4 Schematic diagram of the Tolman and Stewart experiment to measure the charge to mass ratio of the charge carriers.

electrons per turn, we see that the initial kinetic energy of the conduction electrons is $\frac{1}{2}Nm^*pr^2\omega_0^2$. The current I flowing after braking is

$$I = Nev/\pi r = Ne\omega/2\pi \tag{1.27}$$

where ω is the instantaneous angular velocity of the electrons.

When the angular velocity changes from ω to $\omega - \mathrm{d}\omega$, there is an associated change in kinetic energy of $Nm^*pr^2\omega\mathrm{d}\omega$ and this is dissipated by Joule heating. In time $\mathrm{d}t$, this is $I^2R\mathrm{d}t$, where R is the total resistance of the coil and galvanometer circuit. Thus

$$I^2R\mathrm{d}t = Nm^*pr^2\omega\mathrm{d}\omega, \tag{1.28}$$

$$IR\mathrm{d}t = (m^*/e)2\pi pr^2\mathrm{d}\omega. \tag{1.29}$$

Now

$$\int_0^t I\mathrm{d}t = q \tag{1.30}$$

where q is the total charge passed through the galvanometer. Hence

$$\begin{aligned} q &= (m^*/e)(2\pi pr^2/R)\int_0^{\omega_0}\mathrm{d}\omega \\ &= (m^*/e)(Lr\omega_0/R). \end{aligned} \tag{1.31}$$

Table 1.1 *Experimental Lorenz numbers L × 10^8 watt ohm deg^{-2}*

Element	0 °C	100 °C	Element	0 °C	100 °C
Cu	2.23	2.33	Zn	2.31	2.33
Ag	2.31	2.37	Cd	2.42	2.43
Au	2.35	2.40	Pb	2.47	2.56

Thus

$$e/m^* = Lr\omega_0/Rq. \tag{1.32}$$

The right hand side of Equation (1.32) contains parameters which can be determined experimentally. Within the experimental error, Tolman and Stewart found that e/m^* for the current carriers in copper was equal to that for free electrons (as determined by J. J. Thompson's cathode ray experiments). Drude's first assumption was clearly shown to be correct.

1.2.2 Wiedemann–Franz Law

It had, in fact, been noticed earlier in the nineteenth century that the Lorenz number L was a constant for many metals and was almost independent of temperature. A number of examples are cited in Table 1.1. If we use Equation (1.25) to evaluate L we find $L = 1.7 \times 10^{-8}$ watt ohm deg^{-2}. This is not very different from the experimental values listed in Table 1.1. However, the constancy of the Lorenz number is the last success of the Drude model.

1.2.3 Electrical conductivity

Experimentally we find that the resistivity $\rho(=\sigma^{-1})$ for metals varies as

$$\rho = \rho(T) + \rho_0. \tag{1.33}$$

ρ_0 is a temperature independent contribution and varies with specimen perfection and deformation while $\rho(T)$ is a temperature dependent function. At about room temperature $\rho(T)$ varies linearly with absolute temperature T, while at low temperatures $\rho(T)$ varies as T^5. Data for Na and Cu are presented in Fig. 1.5.

According to the Drude theory (Equation (1.12)) if we assume a constant mean free path, we expect $\rho(T)$ to vary as $T^{1/2}$. A clear disparity exists.

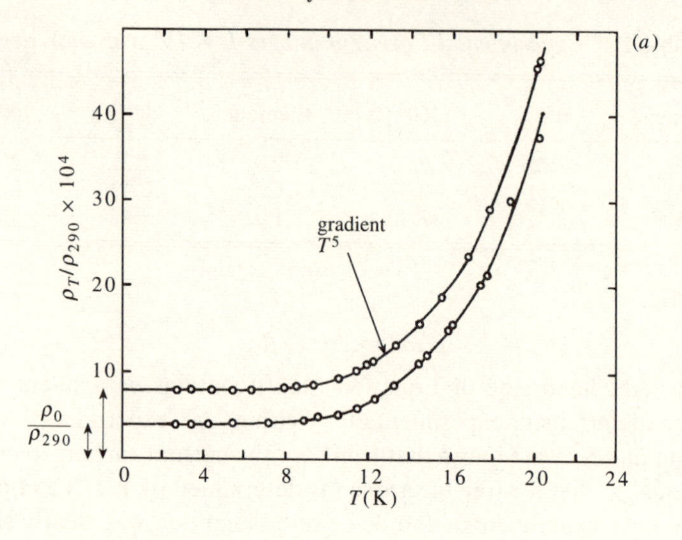

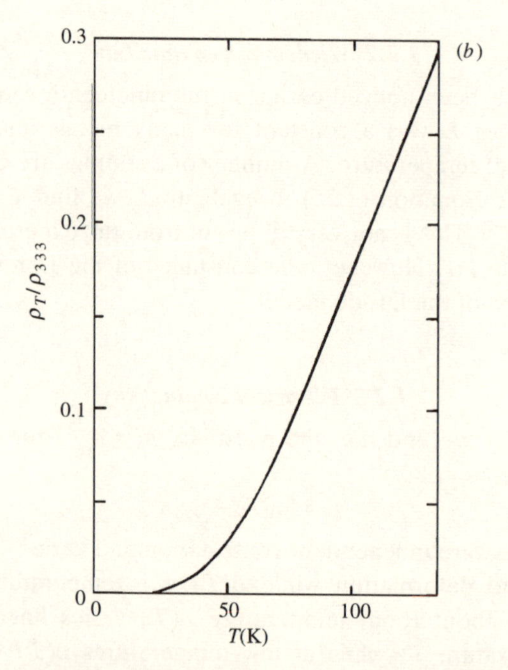

Fig. 1.5 (*a*) Low temperature resistivity ratio of two sodium specimens of different purity. Resistivity ratio is the resistivity at temperature T divided by the resistivity at 290 K. (After D. MacDonald and K. Mendelssohn, Proc. Roy. Soc. (Lond) **A202** (1950) 103. (*b*) Low temperature resistivity ratio (with respect to the Grueisen temperature of 333 K) for copper.

1.2.4 Thermal conductivity

At room temperature, normal metals have thermal conductivities one or two orders of magnitude higher than that of insulators. This supports the supposition that the bulk of the thermal transport in metals occurs via the conduction electrons. However around room temperature the thermal conductivity is almost independent of temperature. According to the Drude theory (Equation (1.19)) we expect a variation as $T^{1/2}$, if we again assume a constant mean free path. Once again the model appears inadequate.

1.2.5 Thermoelectric effects

According to Equation (1.23), the Peltier coefficient should be the same for all metals and hence no heat should be given up or taken in at the junctions (Fig. 1.3). However, we know very well that the Peltier effect is observed and the Peltier coefficients differ in various metals. In fact, the Peltier effect is used for small compact cooling units which are available commercially!

All the variously named thermoelectric effects are related to one another and the Seebeck effect provides a measure of the thermopower. The thermopower S is related to the Peltier coefficient by

$$S = \Pi/T. \tag{1.34}$$

Thus we expect $S \approx 3k_B/2e$ which, on evaluation, is about 100 $\mu V K^{-1}$ and independent of temperature. Experimentally we observe the thermopower to be about 10 $\mu V K^{-1}$ around room temperature and to be proportional to absolute temperature.

1.2.6 Electronic specific heat capacity

At room temperature the specific heat capacity of metals and insulators is very similar and approximately equal to $3n'k_B$ where n' is the number of atoms per unit volume. (This observation is enshrined in Dulong and Petit's Law, that the molar heat capacity is three times the gas constant R. It is straightforward to show that the specific heat capacity of a classical three-dimensional harmonic oscillator is $3k_B$ and by assuming that each atom in the crystal lattice oscillates about its equilibrium position, Dulong and Petit's Law follows at once.) From our viewpoint the most disturbing feature is the absence of a significant contribution due to the conduction electrons in metals. According to

Equation (1.26), and assuming one free conduction electron per atom, we expect the heat capacity of a metal to be 50% higher than that of an insulator. In fact at very low temperatures one can experimentally measure the electronic contribution (Fig. 1.6).

For insulators, lattice dynamics calculations and experimental observations show that at very low temperatures

$$C_v = AT^3. \tag{1.35}$$

However, for metals we observe that

$$C_v = AT^3 + BT, \tag{1.36}$$

where A and B are proportionality constants. The term BT in Equation (1.36) is thus clearly identified as the electronic contribution. In conflict with the predictions of the Drude model, it varies linearly with temperature and at room temperature is about 1/100 of the value predicted by Equation (1.26).

1.3 Discussion and refinements of the classical model

The foregoing sections reveal a very obvious discrepancy between the predictions of the Drude model and experimental observation. While there can be no doubt that the basic assumption of electrons acting as the charge carriers is correct, there must be an important assumption which is invalid. One possibility is the assumption that all electrons have the average energy. A basic axiom of statistical mechanics is that in an assembly of a large number of particles, we can define the total

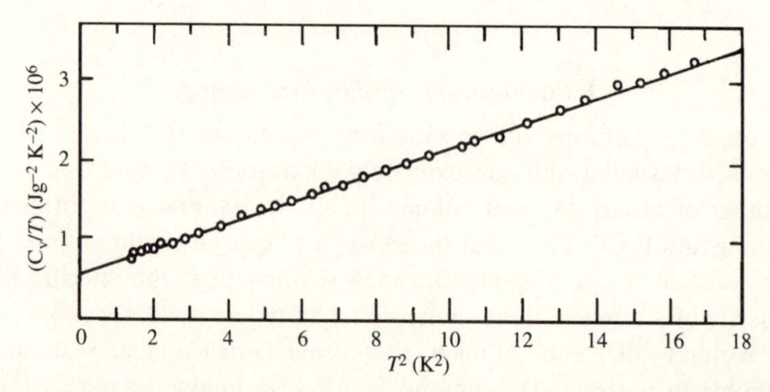

Fig. 1.6 Low temperature specific heat capacity of silver. The non-zero intercept of the C_v/T versus T^2 plot determines the value of the electronic component.

number of particles and the total energy (which is a macroscopically measureable quantity) but we do not know the energies of individual particles. We can (Appendix 2), however, determine a statistical probability distribution for the number of particles within a specified energy region. It is therefore possible to redevelop the theory folding in the energy distribution of the electrons. One can obtain an equation known as the Boltzmann transport equation and by solving this obtain equations for the conductivities and thermopower.

After some quite involved calculations one finds that the form of Equations (1.9), (1.12), (1.17), (1.19), (1.23) and (1.24) do not change. The only difference occurs in the numerical factors – the temperature dependences remain unchanged. Use of the properly weighted distribution does in fact give a very much better value for the Lorenz number which is satisfying. However, the major discrepancies between theoretical and experimental temperature dependences still remain. One further possible invalid assumption is that the mean free path is constant. While we shall see later that we do need a temperature dependent mean free path to explain the room temperature parameters no model can be invoked to give l proportional to $T^{-1/2}$. The problem is much more fundamental.

Worked example: a.c. conductivity

(a) Write down the differential equation for $\langle v \rangle$ in the presence of an alternating field $\mathscr{E}_0 \exp(\mathrm{i}\omega t)$. Look for a solution of the form $\langle v \rangle = \langle v \rangle_0 \exp(\mathrm{i}\omega t)$ and derive an expression for $\langle v \rangle_0$.

(b) Taking only the real part, show that the a.c. conductivity σ is related to the d.c. conductivity σ_0 by

$$\sigma = \sigma_0/(1 + \omega^2\tau^2).$$

(c) Sketch the form of σ as a function of $\omega\tau$.

Solution

(a) In an alternating field, the equation of motion becomes

$$m^* \frac{\mathrm{d}}{\mathrm{d}t}\langle v \rangle + \frac{m^* \langle v \rangle}{\tau} = e\mathscr{E}_0 \mathrm{e}^{\mathrm{i}\omega t}.$$

If we look for a solution $\langle v \rangle = \langle v \rangle_0 \exp(\mathrm{i}\omega t)$ we have on substitution

$$\left(\mathrm{i}\omega m^* + \frac{m^*}{\tau}\right)\langle v \rangle_0 \mathrm{e}^{\mathrm{i}\omega t} = e\mathscr{E}_0 \mathrm{e}^{\mathrm{i}\omega t}.$$

Thus

$$\begin{aligned}\langle v\rangle_0 &= e\mathscr{E}_0/m^*(1/\tau + \mathrm{i}\omega)\\ &= e\mathscr{E}_0\tau/m^*(1 + \mathrm{i}\omega\tau)\\ &= e\mathscr{E}_0\tau(1 - \mathrm{i}\omega\tau)/m^*(1 + \omega^2\tau^2).\end{aligned}$$

(b) If we now take the real part only we have

$$\langle v\rangle_0^{\mathrm{r}} = e\mathscr{E}_0\tau/m^*(1 + \omega^2\tau^2).$$

As $\sigma_0 = ne^2\tau/m^*$ and

$$\sigma = J/\mathscr{E} = ne\langle v\rangle_0^{\mathrm{r}}\mathrm{e}^{\mathrm{i}\omega t}/\mathscr{E}_0\mathrm{e}^{\mathrm{i}\omega t}.$$

Then

$$\sigma = \sigma_0/(1 + \omega^2\tau^2).$$

This is a form of the very common Debye equation, which repeatedly appears in a number of branches of physics.

(c) It is clear from the above equation that as the frequency ω becomes very large, σ falls to zero. Thus we have the form of σ shown in Fig. 1.7.

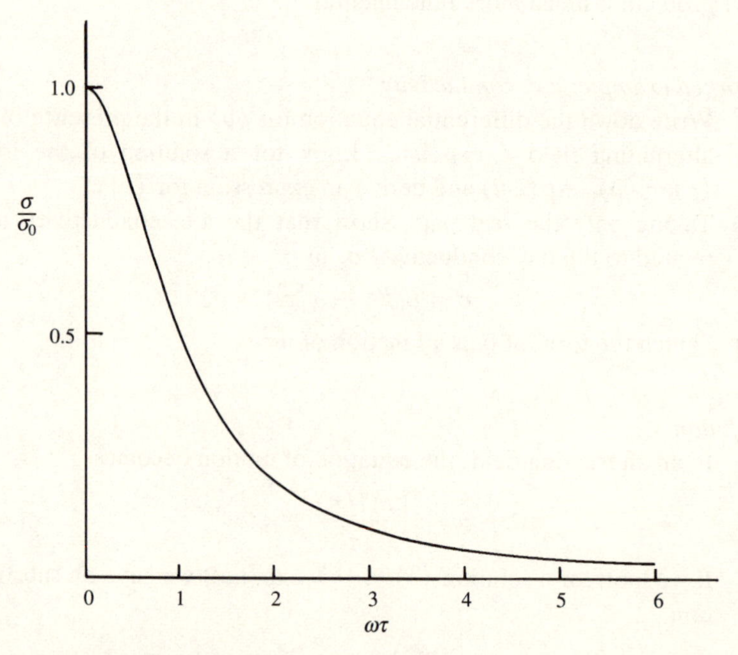

Fig. 1.7 Conductivity of an idealized metal as a function of the product of frequency and relaxation time.

Problems

1.1 A circular coil of wire consisting of N turns of radius r is suspended by a thin fibre so that its plane is horizontal and forms a torsional pendulum with very little damping. If the coil carries a current I show that the angular momentum Γ carried out by the current carriers of mass m and charge e is

$$\Gamma = 2\pi r^2 NIm/e.$$

(Hint: Relate the current to the number and velocity of the carriers using the definition of current.)

When the current is rapidly reversed an impulse of 2Γ is imparted to the coil. If the current reversal occurs as the swinging coil passes through its equilibrium point, show that the amplitude of oscillation changes by $\Delta\theta$, given by

$$\Delta\theta = 2\Gamma(T/2\pi J)$$

where T is the period of oscillation and J the moment of inertia of the coil. In the original paper of Kettering and Scott (*Phys. Rev.* **66** (1944) 257) the average value of r was typically 2 cm, N was 5000, I was 10 mA, T was 60 sec, and J was typically 300 g cm^{-2}. For these values calculate $\Delta\theta$ if $m/e = 5.69 \times 10^{-9}$ g C^{-1}.

1.2 (a) Calculate how many conduction electrons pass through a given area per second when a current of 1 A flows.
(b) Estimate the thermal velocity of an electron in a metal at 300 K.
(c) Estimate the drift velocity of an electron in a current of 50 A in a wire 1 mm square, made from a metal of conductivity 10^8 ohm^{-1} m^{-1}.

1.3 A constant current is passed through a piece of copper wire for an extended period and then the circuit is abruptly broken (e.g. by throwing a switch). Determine the time taken for the current to decay to one half its value if the inductance and capacitance of the circuit are effectively zero.

(The conduction electron concentration in copper is 8.45×10^{28} m^{-3}, $e = 1.6 \times 10^{-19}$ C and $m = 9 \times 10^{-31}$ kg.)

1.4 Integrate Equation (1.21). What do you conclude if $\langle v \rangle/D$ is very small? Is this approximation physically reasonable?

1.5 The electrical resistivity of a sample of pure copper increases by 30% when the temperature is raised from 270 K to 370 K. On addition of nickel to the copper, the resistivity rises by 1.25×10^{-8} ohm m per atomic % Ni. How much nickel must be added so that the resistivity of the alloy increases by only 6% over the same temperature interval? (The resistivity of pure copper is 1.6×10^{-8} ohm m at 270 K.)

2

Quantum mechanical free electron model

The years of the third decade of the present century were heady times. Old social orders were being swept away, monarchies becoming republics and the United States of America was emerging as a world power. In physics too, revolutions were taking place, none more potent than that which resulted in the emergence of quantum mechanics as a model for the behaviour of sub-atomic particles.

It was Newton who first suggested that light was particulate but during the nineteenth century this view had fallen into neglect. Indeed a number of experiments, for example the classic Young's slits experiment, demonstrated quite conclusively that light was a wave motion. How else could interference fringes occur? However, the experiments on the photoelectric effect demonstrated just as conclusively that light energy was carried in packets, or quanta, and that a continuous wave description was not applicable.

This 'wave or particle' dilemma was not unique to the behaviour of light. J. J. Thompson showed that cathode rays were charged, had a well defined mass and had all the properties expected of a beam of particles. Nevertheless, Davisson and Germer showed that diffraction of electrons could take place, an effect made visually much more dramatic by G. P. Thompson's transmission electron diffraction patterns. Nowadays, electron diffraction is used as a routine analytical tool in all advanced metallurgical and materials science laboratories (Fig. 2.1). Clearly new axioms were required.

The new quantum mechanics was enunciated in different forms by Heisenberg and Schrödinger. Schrödinger used the ideas of de Broglie that each particle had an associated wavefunction to develop a detailed wave mechanics. Starting from de Broglie's equations

$$\lambda = h/p \tag{2.1}$$

Fig. 2.1 Transmission electron diffraction pattern from a thin single crystal film of CdS. Electron beam along the body diagonal, hence the hexagonal symmetry. (Courtesy Dr K. Durose, Durham University.)

$$\omega = E/\hbar \tag{2.2}$$

($h = 2\pi\hbar$ is Planck's constant) which link the angular frequency ω and wavelength λ of the 'pilot waves' with the momentum p and energy E of the particle, one can see that the classical expression for total energy of a particle of mass m^*

$$E = (p^2/2m^*) + V, \tag{2.3}$$

becomes

$$\hbar\omega = (h^2/2m^*\lambda^2) + V \tag{2.4}$$

where V is the potential energy.

The magnitude of the wavevector $\mathbf{k}$ is given by

$$k = 2\pi/\lambda \tag{2.5}$$

and for a plane wave at point r and time t of the form

$$\psi(\mathbf{r}, t) = \exp[\mathrm{i}(\mathbf{k}\cdot\mathbf{r} - \omega t)] \tag{2.6}$$

Equation (2.4) becomes

$$(\hbar^2 k^2/2m^*) + V = \hbar\omega. \tag{2.7}$$

The equation for the wavefunction must be compatible with both (2.6) and (2.7), and it is readily seen that

$$\frac{-\hbar^2}{2m^*}\nabla^2\Psi(\mathbf{r}, t) + V(\mathbf{r}, t)\Psi(\mathbf{r}, t) = \mathrm{i}\hbar\frac{\partial}{\partial t}[\Psi(\mathbf{r}, t)] \tag{2.8}$$

satisfies these criteria. This is the famous time dependent Schrödinger equation (where, in Cartesian coordinates $\nabla^2 = \partial^2/\partial x^2 + \partial^2/\partial y^2 + \partial^2/\partial z^2$.)

If the potential energy V is a function of position only, i.e. is time independent, then we can separate the wavefunction $\Psi(\mathbf{r}, t)$ into two parts;

$$\Psi(\mathbf{r}, t) = \psi(\mathbf{r})\Phi(t). \tag{2.9}$$

From Equation (2.8) we have

$$\Phi(t) = \exp(-\mathrm{i}\,Et/\hbar) \tag{2.10}$$

and the time independent Schrödinger equation

$$\frac{-\hbar^2}{2m^*}\nabla^2\psi(\mathbf{r}) + V\psi(\mathbf{r}) = E\psi(\mathbf{r}). \tag{2.11}$$

It is this equation which we must use to understand the behaviour of electrons in solids.

2.1 Sommerfeld free electron model

The model developed by Sommerfeld was really very similar to that which we considered in Chapter 1, except that there the electrons were not constrained by the postulates of quantum mechanics. The electrons were considered to be free to move randomly within a flat bottomed potential well, whose dimensions were that of the sample and which could be considered to be of infinite depth for purposes of calculation. The constraint that Schrödinger's equation must be satisfied for this potential well imposes restrictions on the allowed energies which an electron can take up. We will first examine the allowed energies in one dimension.

For the model potential well of width L shown in Fig. 2.2 we have

$$\frac{-\hbar^2}{2m^*}\left(\frac{\mathrm{d}^2\psi}{\mathrm{d}x^2}\right) = E\psi \quad \text{for } 0 < x < L. \tag{2.12}$$

No allowed solutions exist outside the range $0 < x < L$ and ψ must be zero at both $x = 0$ and $x = L$. A possible solution of Equation (2.12) is

$$\psi = A \sin kx + B \cos kx \tag{2.13}$$

where A and B are proportionality constants.

As $\psi = 0$ at $x = 0$ then $B = 0$. Also $\psi = 0$ at $x = L$ and thus either $A = 0$ (a trivial solution) or $\sin kL = 0$. This implies

$$k = n_x\pi/L \tag{2.14}$$

where n_x is an integer.

Substituting (2.14) into (2.12) and (2.13) yields

$$E = h^2 n_x^2/8m^* L^2. \tag{2.15}$$

This gives the allowed energy levels of an electron within the solid. The first few are sketched in Fig. 2.2. We note that because L is of macroscopic dimensions, adjacent energy levels are very close together and can be considered as a quasi-continuum. The density of states $D(E)\,\mathrm{d}E$ is defined as the number of states between energies E and $E + \mathrm{d}E$. It is twice the density of levels defined by (2.15) because electrons have spin and the spin can have two values. In zero applied magnetic field the energies of these two spin states are equal. Thus for every quantum number n_x, there are two electron states.

2.1.1 Pauli exclusion principle

So far we have discovered the allowed energy levels for a one-dimensional metal such as a conducting polymer but have no information

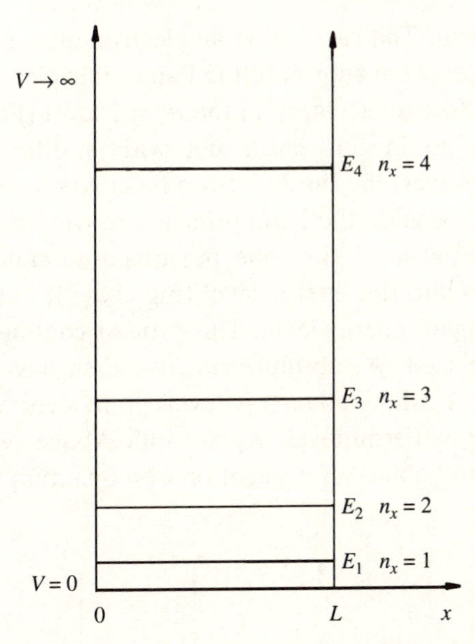

Fig. 2.2 Quantum states in a one-dimensional, infinite depth, potential well.

about how these levels are populated. Classically, the probability $P(E)$ that a particle (in a large ensemble) will have energy E is given by the Boltzmann distribution, i.e.

$$P(E) = \xi \exp(-E/k_B T), \tag{2.16}$$

where ξ is a normalizing constant. This implies that at absolute zero, all particles will have zero energy. As the temperature rises so a higher energy becomes more probable, but still the most probable energy is zero. The average energy is zero at $T = 0$ and at temperature T, for a one-dimensional ensemble, is given by

$$\langle E \rangle = k_B T/2. \tag{2.17}$$

While Boltzmann statistics are appropriate to our classical model with *distinguishable* particles any member of which can have a given energy, it is not applicable to the quantum mechanical model. Electrons are subject to the Pauli exclusion principle and this means that no two electrons may occupy the same quantum state. Consider now the hypothetical situation in which we remove all the (free) conduction electrons, one each for each atom. (For the moment we will ignore the fact that the solid would not then hold together!) We will proceed to return these one at a time and see in what energy level subsequent electrons are placed. The rule is that an electron must be in the lowest available energy level but subject to the Pauli principle.

Thus the first electron is placed in the $n_x = 1$ level (Fig. 2.3(*a*)). The second can also go in this level, but with a different spin state (Fig. 2.3(*b*)). However, the third electron must go into a higher energy level (Fig. 2.3(*c*)) because the Pauli principle restricts the filling to two electrons per energy level (i.e. one per quantum state). The fourth electron also goes into this energy level (Fig. 2.3(*d*)) but the fifth must again go into a higher energy level. This process continues until all the electrons are replaced. At absolute zero we then have the situation, sketched in Fig. 2.4, where all energy levels up to a certain value called the Fermi energy or Fermi level, E_F are full. Above E_F all the levels are empty. The probability of occupation of a quantum state, $P(E)$, is

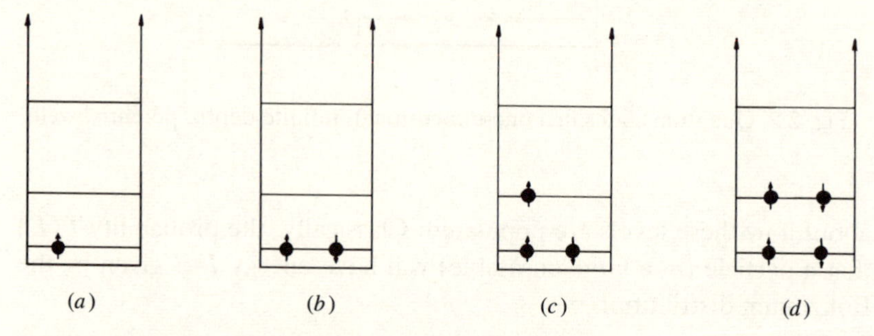

Fig. 2.3 Schematic representation of the filling of quantum states from the lowest energy, subject to the Pauli exclusion principle.

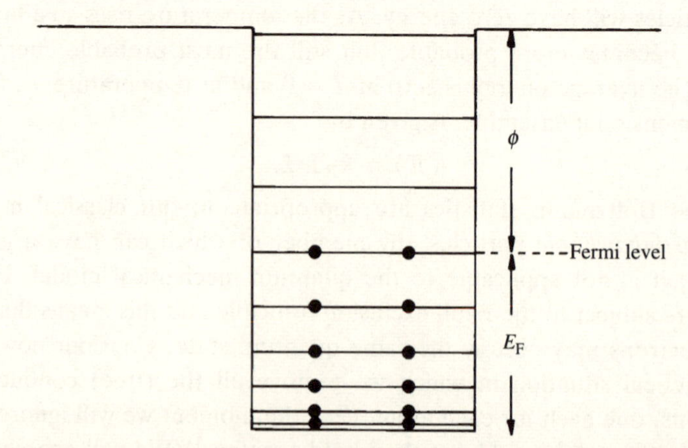

Fig. 2.4 Schematic representation at absolute zero of the 'Fermi sea' of filled electron states up to a Fermi energy E_F. The work function is shown as ϕ.

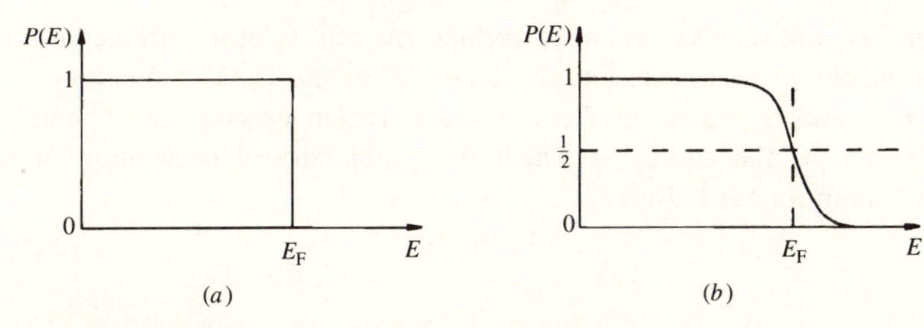

Fig. 2.5 The Fermi–Dirac distribution (*a*) at absolute zero (*b*) at non-zero temperature.

thus as shown in Fig. 2.5(*a*). At non-zero temperature, some excitation to higher, unoccupied levels takes place. However, only a small fraction of electrons have vacant states within an energy $k_B T$, the average energy available thermally and these are the electrons close to the Fermi level E_F. Remembering that vacant states below E_F must have an equivalent occupied state above E_F, it is quite easy to understand how the distribution shown in Fig. 2.5(*b*)) arises. As the temperature rises, so the tail of the distribution becomes longer. Mathematically, the distribution is expressed as

$$P(E) = \{[\exp(E - E_F)/k_B T] + 1\}^{-1} \tag{2.18}$$

and this is known as the Fermi–Dirac distribution. It holds, quite generally for indistinguishable particles of half integral spin, i.e. those governed by the Pauli exclusion principle.

2.1.2 Electron density

Using Fermi–Dirac statistics and appropriate solutions of Schrödinger's equation we are now in a position to determine the distribution of electrons with energy. The electron density $N(E)\,dE$ between energies E and $E + dE$ is the product of two totally independent quantities, the density of states $D(E)\,dE$ between E and $E + dE$, and the probability of occupation of a quantum state. Mathematically we have

$$N(E)\,dE = P(E)D(E)\,dE. \tag{2.19}$$

It is important at this stage to emphasize that $P(E)$ is totally independent of the density of states and arises from statistical mechanical considerations only. The probability of occupation of a state is not influenced by whether there is a state there to be occupied! Thus, later on, we will see that when we include the effects of a periodic lattice some energy ranges have a zero density of states. The Fermi energy E_F can however still lie in that forbidden region because E_F is simply defined as the energy at which the probability of occupation of a quantum state is $\frac{1}{2}$. Thus,

$$P(E_F) = \tfrac{1}{2}. \tag{2.20}$$

2.1.3 Density of states in one dimension

In one dimension we see from Equation (2.14) that there is one value of wavenumber k for every interval π/L. This can be alternatively stated as the density of states $D(k)\,dk$ between k and $k + dk$ given by

$$D(k)\,dk = \frac{L}{\pi}\,dk. \tag{2.21}$$

This can be converted to an energy scale by noting that the number of states $D(k)\,dk$ between k and dk is *equal* to the number of states $D(E)\,dE$ between equivalent energies E and $E + dE$. Thus,

$$D(E)\,dE = D(k)\,dk = (L/\pi)\,dk. \tag{2.22}$$

As

$$k = (2m^* E/\hbar^2)^{1/2} \tag{2.23}$$

we have

$$D(E)\,dE = \left(\frac{L}{\pi}\right)\left(\frac{m^*}{2\hbar^2 E}\right)^{1/2} dE. \tag{2.24}$$

This variation as $E^{-1/2}$ is characteristic of a one-dimensional system.

At absolute zero, the Fermi energy E_F corresponds to the energy of the top-most filled level. We can determine it by integration of Equation (2.24). As $P(E) = 1$ for $E < E_F$ and $P(E) = 0$ for $E > E_F$ the electron density at $T = 0$ is zero for $E > E_F$ and equal to $D(E)\,dE$ for $E < E_F$. The total number of electrons N is given by

$$N = \int_0^{E_F} 2\left(\frac{L}{\pi}\right)\left(\frac{m^*}{2\hbar^2 E}\right)^{1/2} dE \tag{2.25}$$

where we have now introduced a factor of 2 to account for the two spin states. Thus

$$E_F = \frac{h^2}{32 m^*}\left(\frac{N}{L}\right)^2. \tag{2.26}$$

a result which can also be found by differentiation of Equation (2.15).

We can obtain an estimate of the magnitude of E_F on the assumption of one free electron per atom, spaced 1 Å apart. This yields a value of E_F of the order of 1 eV. Note that this is very high compared with thermal energy at room temperature. (All physicists should remember that $k_B T \approx 1/40$ eV at room temperature!) From the one-dimensional density of states (Equation (2.25) we can also determine the average electron energy $\langle E \rangle$ at $T = 0$. This will be

$$\langle E \rangle = \frac{\int_0^{E_F} E \cdot E^{-1/2} \, dE}{\int_0^{E_F} E^{-1/2} \, dE} = E_F/3. \tag{2.27}$$

This is also very high compared with the average energy of a classical particle, namely $k_B T/2$ (Equation (2.17)).

2.1.4 Density of states in three dimensions

Only a few materials, such as the sulphur–nitrogen polymer $(SN)_x$ display approximately one-dimensional metallic properties. The common metals are remarkably isotropic in their properties and we must therefore derive the density of states in three dimensions, not one. To do this we need to solve the Schrödinger equation again, but this time in the full three-dimensional form.

A general standing wave solution of the time independent Schrödinger equation,

$$\frac{-\hbar^2}{2m^*}\left(\frac{\partial^2 \psi}{\partial x^2} + \frac{\partial^2 \psi}{\partial y^2} + \frac{\partial^2 \psi}{\partial z^2}\right) = (E - V)\psi, \tag{2.28}$$

is

$$\begin{aligned}\psi = {} & A \sin k_x x \sin k_y y \sin k_z z + B \sin k_x x \sin k_y y \cos k_z z \\ & + C \sin k_x x \cos k_y y \sin k_z z + D \sin k_x x \cos k_y y \cos k_z z \\ & + E \cos k_x x \sin k_y y \sin k_z z \\ & + F \cos k_x x \cos k_y y \sin k_z z + G \cos k_x x \sin k_y y \cos k_z z \\ & + H \cos k_x x \cos k_y y \cos k_z z \end{aligned} \tag{2.29}$$

where *A, B, C, D, E, F, G* and *H* are proportionality constants.

Let us now consider the solid to be a cube of side L, with cube edges parallel to the x, y and z axes (Fig. 2.6(a)). Within the cube the potential energy is zero, outside the cube it is infinite. Therefore the

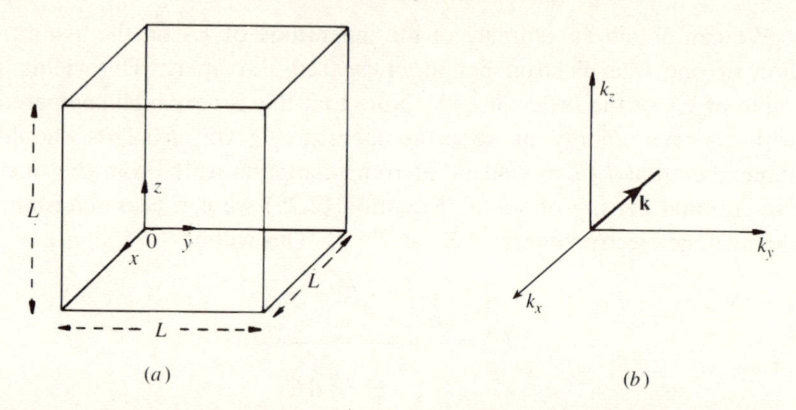

Fig. 2.6 (*a*) Real space coordinate system defining a three-dimensional cubic box with infinite potential at the edges. (*b*) Wavevector space (often called reciprocal or phase space) coordinate system showing the electron wavevector **k**.

wavefunction ψ must be zero at the edges of the cube. Provided that L is large compared with the de Broglie wavelength of the electron this is a good approximation to a finite depth potential well. The requirement that $\psi = 0$ at $x = 0$, $y = 0$ and at $z = 0$ means that all coefficients of Equation (2.29) are zero except A. Setting $\psi = 0$ independently at $x = L$, $y = L$ and $z = L$, yields

$$\left.\begin{aligned} k_x &= n_x\pi/L \\ k_y &= n_y\pi/L \\ k_z &= n_z\pi/L \end{aligned}\right\} \qquad (2.30)$$

where n_x, n_y and n_z are positive integers, k_x, k_y and k_z are the Cartesian components of a vector **k** called the wavevector and related to the components by

$$k^2 = k_x^2 + k_y^2 + k_z^2. \qquad (2.31)$$

Note that wavevector **k** has dimensions of $(\text{length})^{-1}$ (Fig. 2.6(*b*)). Substitution of Equation (2.30) into (2.28) yields the energy eigenvalues E_n,

$$E_n = \left[\frac{\hbar^2}{2m^*}\left(\frac{\pi^2}{L^2}\right)\right](n_x^2 + n_y^2 + n_z^2) \qquad (2.32a)$$

$$E_n = \frac{\hbar^2}{2m^*}(k_x^2 + k_y^2 + k_z^2) \qquad (2.32b)$$

$$E_n = \frac{\hbar^2 k^2}{2m^*}. \qquad (2.32c)$$

Equation (2.32c) is identical to the energy–wavevector relation for a free particle except that the values of k are estimated through Equations (2.30) and (2.31).

The allowed energy values are established by positive integer values of n_x, n_y and n_z. Thus, in a coordinate system n_x, n_y, n_z as shown in Fig. 2.7(*a*) the only allowed energy values are those corresponding to integer coordinates. In this coordinate space, one value of energy occupies a cube of unit volume. Transforming, using Equation (2.30), to the wavevector space (k space) of Fig. 2.7(*b*) we see that each value of energy occupies volume $(\pi/L)^3$ of the positive octant of k space. Over the whole of the wavevector (k) space we have one allowed value in volume $8(\pi/L)^3$. Therefore the number of allowed values between wavevector k and $k + \mathrm{d}k$ is $(L/2\pi)^3$ times the volume of a shell between radii k and $k + \mathrm{d}k$. Thus,

$$D(k)\,\mathrm{d}k = 4\pi\left(\frac{L}{2\pi}\right)^3 k^2\,\mathrm{d}k \tag{2.33}$$

From Equation (2.32c) we have

$$\begin{aligned} D(E)\,\mathrm{d}E &= \frac{(2m^{*3})^{1/2}}{\pi^2\hbar^3} L^3 E^{1/2}\,\mathrm{d}E \\ &= \frac{(2m^{*3})^{1/2}}{\pi^2\hbar^3} V E^{1/2}\,\mathrm{d}E \end{aligned} \tag{2.34}$$

Fig. 2.8(*a*) shows the form of this variation. Unlike the one-dimensional case, there is no singularity at $E = 0$. We have included a factor of 2 to take the two spin states into account and replaced L^3 by the

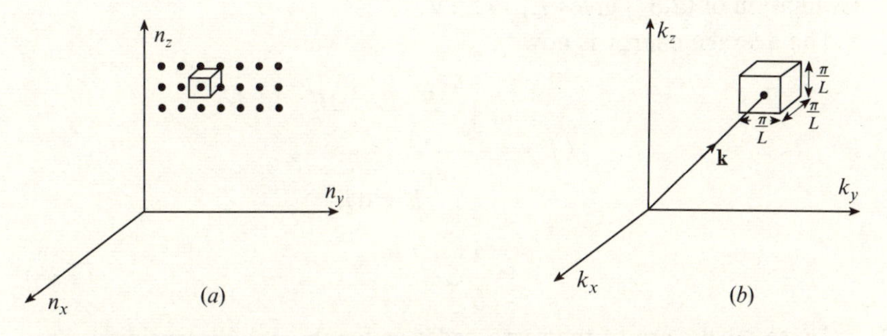

Fig. 2.7 (*a*) Coordinate system defining the allowed quantum states. (*b*) Equivalent diagram in reciprocal space showing the volume containing one allowed value of the wavevector.

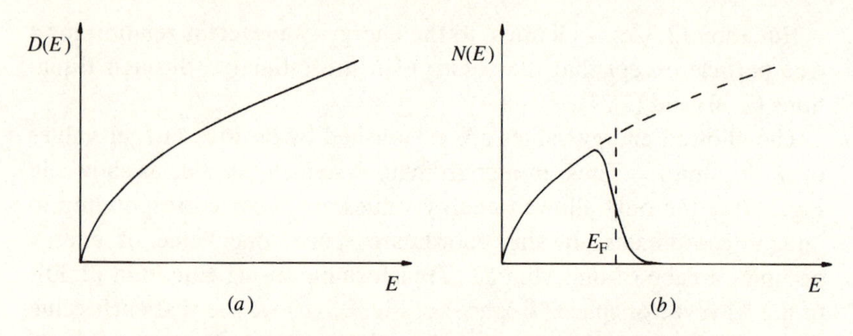

Fig. 2.8 (*a*) Density of states for a three-dimensional free electron gas. (*b*) Electron density as a function of energy at non-zero temperature.

sample volume V. The number of electrons with energies between E and $E + \mathrm{d}E$, is given by

$$\begin{aligned} N(E)\,\mathrm{d}E &= P(E)D(E)\,\mathrm{d}E \\ &= \frac{(2m^{*3})^{1/2}}{\pi^2\hbar^3}V\left\{\frac{E^{1/2}\,\mathrm{d}E}{\exp\left[(E-E_\mathrm{F})/k_\mathrm{B}T\right]+1}\right\}. \end{aligned} \tag{2.35}$$

This is sketched in Fig. 2.8(b).

We can now evaluate the Fermi energy, by integration of Equation (2.35) for $T = 0$. Here, again, $P(E) = 1$ for $E < E_\mathrm{F}$ and $P(E) = 0$ for $E > E_\mathrm{F}$. Explicitly we have

$$N = \int_0^{E_\mathrm{F}} \frac{(2m^{*3})^{1/2}}{\pi^2\hbar^3} V E^{1/2}\,\mathrm{d}E, \tag{2.36}$$

i.e.

$$E_\mathrm{F} = \frac{\hbar^2}{2m^*}\left(\frac{3\pi^2 N}{V}\right)^{2/3}. \tag{2.37}$$

Evaluation of (2.37) gives $E_\mathrm{F} \approx 5$ eV.

The average energy is now

$$\begin{aligned} \langle E\rangle &= \frac{\displaystyle\int_0^{E_\mathrm{F}} E\cdot E^{1/2}\,\mathrm{d}E}{\displaystyle\int_0^{E_\mathrm{F}} E^{1/2}\,\mathrm{d}E} \\ &= \frac{3}{5}E_\mathrm{F}. \end{aligned} \tag{2.38}$$

These results are of the same order as for the one-dimensional case. While the average energy does rise when $T \neq 0$, the effect is small and is in second order in $(k_\mathrm{B}T/E_\mathrm{F})$. Again we see that the effect of the exclusion principle is to push up the average energy by two orders of magnitude.

2.1.5 The Fermi surface

At this point we introduce, as an almost trivial extension of the foregoing discussion, a very important concept which repeatedly occurs throughout the subject of solid state physics. This is the Fermi surface. It is the surface in k space corresponding to a surface of constant energy E_F. In the free electron model the energy depends only on the magnitude, not the direction of the wavevector (see Equation (2.32c)) and the surfaces of constant energy are spheres about the origin. The Fermi surface is thus a sphere (Fig. 2.9).

While simple for the free electron model, this surface becomes distorted when the effects of the periodicity of the crystal lattice are included. Studies of the topology of the Fermi surface are an important and active branch of solid state physics research.

Problems

2.1 If each atom in sodium occupies a volume of 4×10^{-29} m^3 and contributes one free conduction electron, calculate (in eV) the value of the Fermi energy of sodium.

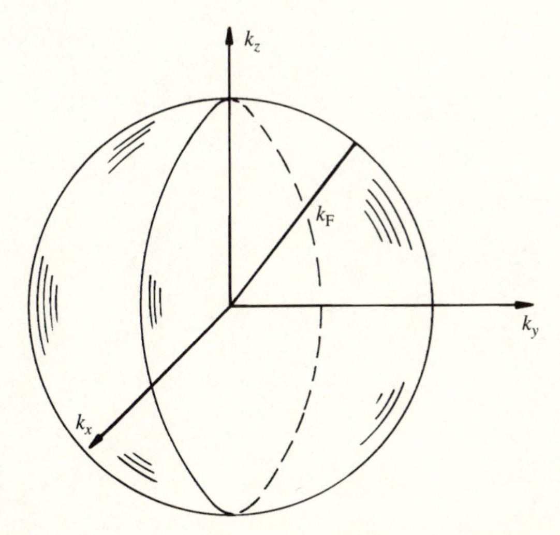

Fig. 2.9 The Fermi surface for a three-dimensional free electron gas.

2.2 The Fermi energy of gold is 5.5 eV. Calculate the velocity of an electron at the Fermi surface, v_F. Compare this with the root mean square velocity calculated using classical kinetic theory.

2.3 Calculate the value of the Fermi temperature T_F (given by $E_F = k_B T_F$) in gold. Compare this with room temperature. Estimate the fractional error introduced into the velocity by assuming all electrons contributing to conduction have velocity v_F.

2.4 Because the atomic bonding is strong within a sheet of atoms, but very weak between sheets, graphite behaves very much like a two-dimensional metal. By solving the Schrödinger equation in two dimensions, prove that the density of states in a two-dimensional metal is

$$D(E) = mL^2/\hbar^2\pi$$

and that the Fermi energy is $n\hbar^2\pi/m^*$ where n is the number of conduction electrons per unit area.

2.5 Prove that the average velocity $\bar{v}$ of electrons in a one-dimensional Fermi gas is $(E_F/2m^*)^{1/2}$.

2.6 Prove that the average energy at absolute zero of a two-dimensional Fermi gas is $E_F/2$.

2.7 Determine the root mean square velocity of a three-dimensional Fermi gas in terms of E_F and m^*.

3

Application of the Fermi gas model

3.1 Thermal properties

3.1.1 Electronic specific heat

The results derived in the previous chapter have shown that only a very small fraction of electrons can be excited thermally. The problem of the overlarge classical electronic specific heat now becomes clear. We have been considering too many electrons. An approximate value of the electronic specific heat can be obtained by inspection of Fig. 2.5(*b*). There it can be seen that the area of the 'tail' above E_F is about $k_B T$. Thus, the fraction of electrons in the thermally excited tail is about $k_B T/E_F$, corresponding to only 1% at room temperature. The classical result must be reduced by a factor of $k_B T/E_F$.

Hence the electronic specific heat, C_e, is given by

$$\begin{aligned} C_e &\approx \frac{3}{2} n k_B \cdot k_B T/E_F \\ &\approx \frac{3}{2} n k_B^2 T/E_F. \end{aligned} \tag{3.1}$$

The electronic contribution to the specific heat therefore varies linearly with temperature and at room temperature is only of the order of 1/100th of the lattice specific heat. This explains why the specific heats of metals and insulators are approximately the same at room temperature. Further, the magnitude of the electronic term, which is best measured at low temperatures where it is comparable to the contribution from the crystal lattice, is of the order predicted by Equation (3.1).

The above argument gives only an approximate value for the slope of C_e versus T. To determine it accurately we must integrate over the density of states distribution. The total change in electron energy ΔE

on heating from absolute zero to a temperature T is given by

$$\Delta E = \int_{E_F}^{\infty} (E - E_F)P(E)D(E)\mathrm{d}E + \int_0^{E_F} (E_F - E)[1 - P(E)]D(E)\mathrm{d}E. \qquad (3.2)$$

Here, the first integral is the change in energy occurring when electrons are taken from E_F to states above E_F and the second represents the change occurring when electrons are taken from states below E_F to E_F. The term $P(E)$ is given by $[\exp(E - E_F)/k_B T + 1]^{-1}$, and $1 - P(E)$ is the probability of an electron *not* being in a state with energy E.

The electronic specific heat is given by

$$C_e = V^{-1}\partial(\Delta E)/\partial T = V^{-1}\int_0^{\infty} (E - E_F)\left[\frac{\partial}{\partial T}P(E)D(E)\right]\mathrm{d}E. \qquad (3.3)$$

For $k_B T \ll E_F$ (i.e. all normal temperatures), $P(E)$ only changes appreciably with temperature close to the Fermi level. We can take the density of states $D(E)$ equal to its value $D(E_F)$ at the Fermi level and outside the integral.

$$\frac{\partial}{\partial T}P(E) = \frac{E - E_F}{k_B T^2}\left(\frac{\exp[(E - E_F)/k_B T]}{\{\exp[(E - E_F)/k_B T] + 1\}^2}\right). \qquad (3.4)$$

Writing $x = (E - E_F)/k_B T$, we have

$$C_e = V^{-1}k_B^2 T D(E_F)\int_{-E_F/k_B T}^{\infty} \frac{x^2 e^x \mathrm{d}x}{(e^x + 1)^2}. \qquad (3.5)$$

The lower limit may be approximated as $-\infty$ and as the integral becomes

$$\int_{-\infty}^{\infty} \frac{x^2 e^x \mathrm{d}x}{(e^x + 1)^2} = \frac{\pi^2}{3}, \qquad (3.6)$$

we have

$$C_e = \frac{\pi^2}{3V}k_B^2 T D(E_F). \qquad (3.7)$$

The density of states at the Fermi level $D(E_F)$ is an important parameter, as we shall see in the next chapter when the effects of a periodic crystal structure are included. It determines whether a solid is an insulator, a semiconductor or a metal. In the free electron model we have, from Equations (2.34) and (2.37),

$$D(E_F) = 3N/2E_F \qquad (3.8)$$

and hence the electronic specific heat capacity is given by

$$C_e = \frac{\pi^2}{2} n k_B^2 T / E_F. \tag{3.9}$$

This result differs from the result of Equation (3.1) only in a constant factor $\pi^2/3$.

Equation (3.7) is much more general than (3.9) and enables us to determine the density of states at the Fermi surface from experimental specific heat measurements. Such measurements still form an active branch of research, particularly in compounds of rare earths and actinides.

3.2 Transport properties

We are now in a position to re-examine the predictions concerning the transport properties. Here we shall see that the effect of the exclusion principle is to reduce the number of electrons that can be scattered to different energy states. It is the average velocity of *these* electrons which determines the momentum loss and hence the resistivity.

3.2.1 Electrical conductivity

When an electric field is applied across the ends of a sample, all electrons experience the Coulomb force. Acceleration of the whole electron distribution therefore occurs and, as illustrated in Fig. 3.1, the whole Fermi surface moves to the right. The current density J is still defined by Equation (1.6). However, the drift velocity is not the same as for the classical model. Only electrons within about $k_B T$ of the Fermi surface can see vacant energy states into which they can be scattered. An example, giving a small change in energy, but a large change in drift momentum is the transition from A to B in Fig. 3.1.

As these are the only electrons which can be scattered, the average velocity of the scattered electrons, i.e. those responsible for the destruction of gained momentum, is not the classical value of velocity, but v_F, the velocity of an electron at the Fermi surface. This is given simply by

$$E_F = \frac{1}{2} m^* v_F^2. \tag{3.10}$$

Thus the mean free path, l, is given by

$$l = v_F \tau \tag{3.11}$$

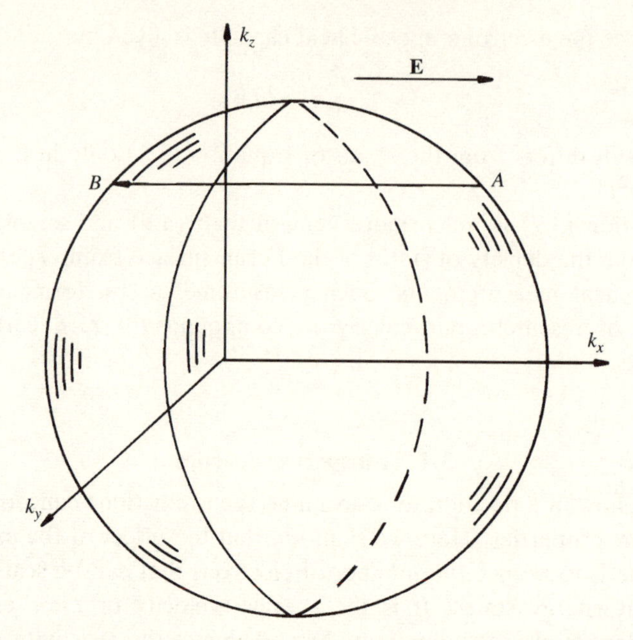

Fig. 3.1 Shift in the electron density distribution as a function of wavevector on application of an electric field, **E**.

and hence from Equation (1.9) the conductivity, σ, is

$$\sigma = ne^2l/m^*v_F. \tag{3.12}$$

3.2.2 Thermal conductivity

The argument used above for electrical conductivity holds identically for the thermal conductivity. Therefore we have immediately from Equations (1.17) and (3.11)

$$K = 9k_B^2nlT/4m^*v_F. \tag{3.13}$$

3.2.3 Scattering mechanisms

In order to compare the predictions of Equations (3.12) and (3.13) with experiment we must consider the scattering mechanisms that occur and first we must identify the obstacles with which the electrons collide. An immediate guess might be the ions making up the crystal lattice. However, as the spacing between ions or atoms in all solids is of the order of 1 Å (10^{-10} m), the mean free path would be of similar magnitude. Further, because the thermal expansion coefficients are only a few parts in 10^6, l would not vary appreciably with temperature. Experimentally we find $l \approx 100$ Å (10^{-8} m) at room temperature, rising to above a millimetre for pure solids at low temperature.

While not apparent in the free electron model, it turns out that the conduction electrons are not scattered by the ions in a perfectly periodic lattice. (Energy band theory explains this in terms of the electron wavefunction taking up the periodicity of the lattice.) However, the electrons *are* scattered if the lattice is not perfectly periodic. This departure from periodicity can arise in two ways.

The first is that the lattice may be imperfect. A real lattice may, and indeed does, contain impurity atoms which do not fit properly into the host lattice. It will also have a certain number of vacant sites, lattice sites where no ion is present. It may also have linear defects known as dislocations or planar defects such as stacking faults, boundaries where the sequential stacking of atomic planes is interrupted. Polycrystalline materials (such as the metals in everyday use) will have many boundaries between the small crystallites oriented in different directions. (It is often not appreciated that materials such as mild steel are composed of many small crystals, typically 10^{-5} m in diameter, oriented in almost random directions. A macroscopic example can be seen on fresh galavanized iron, where the zinc crystals coating the iron have different reflectivities to light, depending on their orientation. They are easily visible with the naked eye.) At low temperatures these crystal defects provide the dominant scattering mechanism. The mean free path is almost independent of temperature and varies with the perfection of the specimen. It therefore provides a temperature independent contribution to the electrical resistivity ρ_0. This is indeed seen experimentally.

A temperature independent l can be seen, from Equation (3.11), to give rise to a low temperature thermal conductivity linear with absolute temperature T. This is again in good agreement with the low temperature variation of the thermal conductivity of metals.

The second departure from periodicity occurs as the temperature rises and the crystal lattice begins to vibrate. (We refer to solids as being 'hot' when they have a large amount of this vibrational energy.) Such vibration means that the conduction electrons will very often experience ions which are a long way from their equilibrium positions and this causes scattering of the conduction electrons. It is worth noting that the lattice vibration is a collective effect because neighbouring ions interact strongly. These collective vibrations must be zero at the specimen boundary and, for exactly the same boundary condition arguments used in Chapter 2, solution of the wave equation for the elastic wave leads to quantization of the lattice vibrations. The quanta of elastic energy can be considered as quasi-particles and these are known as phonons. It is convenient to consider the scattering of electrons by the vibrating lattice in terms of particle–particle collisions between the electrons and phonons.

Consideration of lattice dynamics can show that the phonon density varies linearly with temperature at around room temperature. This implies that the mean free path varies inversely with temperature. From Equation (3.12) we can see that this implies that electrical resistivity varies linearly with temperature at around room temperature. This is what is observed experimentally, as here the thermal scattering dominates. The two types of scattering mechanism, defect and thermal, account for Matthiesen's Law, i.e. the resistivity is the sum of a temperature independent part and a temperature dependent part ($\rho = \rho_0 + \rho(T)$).

Substitution of a mean free path inversely proportional to temperature into Equation (3.13) yields a thermal conductivity independent of temperature at around room temperature. This is also in good agreement with experiment.

3.2.4 Thermoelectric effects

As we saw in Chapter 1, the classical free electron theory predicts a value of the Peltier coefficient which is much higher than that observed experimentally. Again we have counted too many electrons, but the argument is more subtle than before. An initial glance at Equation (1.23) suggests that scaling should occur just as in the Wiedemann–Franz Law (Equation (1.25)) and that the Drude and Sommerfeld models should give the same result.

However, although both electric charge and kinetic energy are carried by all the electrons, only those within approximately $k_B T$ of the Fermi energy can lose or gain energy at the junction. While the whole Fermi surface is shifted (Fig. 3.1) only a fraction ($k_B T/E_F$) of the kinetic energy can appear as thermal (free) energy because only this fraction of the conduction electrons have vacant states into which they can be excited. Accordingly, the heat given out or taken up at the junction is reduced by a factor of $k_B T/E_F$. Thus, the Peltier coefficient becomes

$$\Pi = 3(k_B T)^2/2eE_F, \tag{3.14}$$

and the thermopower S is given by

$$S = 3k_B^2 T/2eE_F. \tag{3.15}$$

The dependence of S on absolute temperature is indeed observed experimentally.

3.3 Magnetic properties

3.3.1 Pauli paramagnetism

In the free electron model, the energy of an electron does not depend on its spin state if the magnetic field is exactly zero. However, in an applied magnetic field, the energies of the two spin states are different and some very interesting magnetic phenomena are observed as a consequence. Associated with an electron is an angular momentum called spin angular momentum and with this angular momentum is associated a magnetic moment. A useful pictorial model, but one that should be used with caution, is of an electron spinning like a child's top about an axis. From spectroscopic measurements we know that the electron has a magnetic dipole moment $\boldsymbol{\mu}$ given by

$$\boldsymbol{\mu} = -g_s \mu_B \mathbf{S} \tag{3.16}$$

here $g_s = 2$, μ_B is the Bohr magneton and $\mathbf{S}$ is the angular momentum associated with the electron spin in units of $\hbar$. In a magnetic field of flux density $\mathbf{B}$, the magnetic dipole will possess an energy U associated with its orientation with respect to the field direction, given by

$$\begin{aligned} U &= -\boldsymbol{\mu} \cdot \mathbf{B} \\ &= -\mu B \cos\theta \end{aligned} \tag{3.17}$$

where θ is the angle between $\boldsymbol{\mu}$ and $\mathbf{B}$. The lowest energy state is therefore when the dipole moment is aligned parallel to the field direction.

It was the, now classic, experiment by Stern and Gerlach on an atomic beam of silver atoms which first demonstrated that there was a spatial quantization of such a system. Their experiment showed that only two orientations of the dipole with respect to the field were allowed. These correspond to values of the component of angular momentum in the field direction of $\pm\hbar/2$ (Fig. 3.2). Thus the energy of

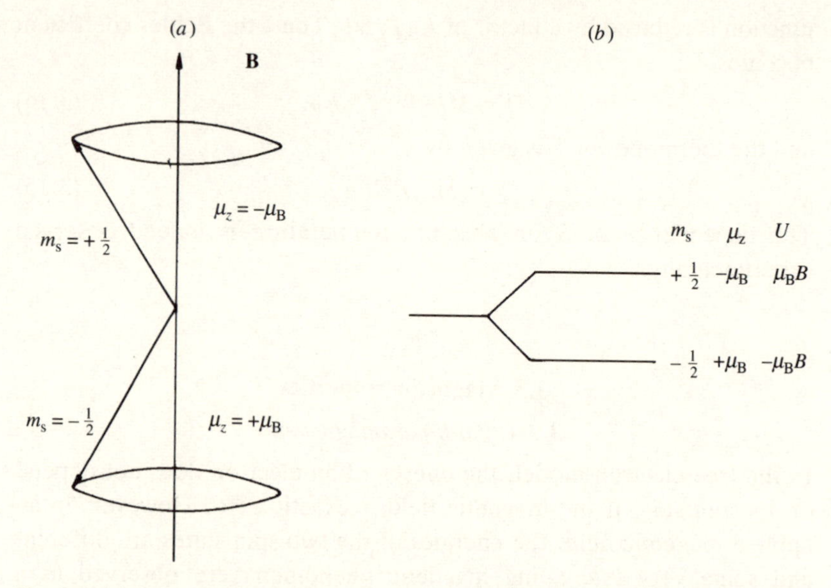

Fig. 3.2 (*a*) Spin angular momentum and *z* component of magnetic moment of an electron in a magnetic field **B**. (*b*) Equivalent energy levels showing the splitting from the zero field situation.

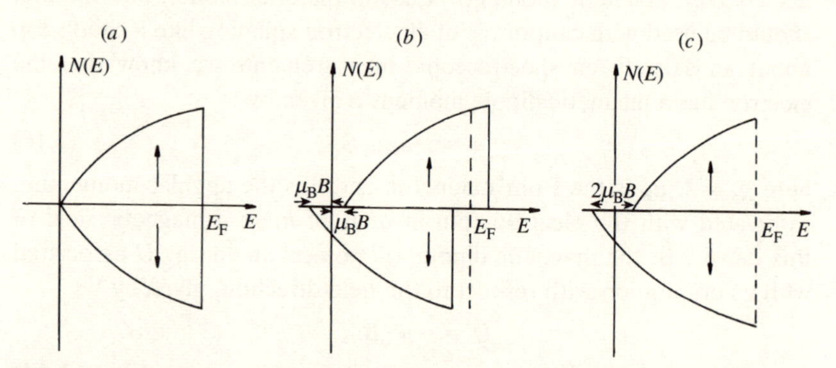

Fig. 3.3 (*a*) Spin-up and spin-down distribution in zero field. (*b*) Non-equilibrium distributions on application of field **B**. (*c*) Equilibrium distributions showing transfer of electrons between spin states.

the electron associated with its orientation in the magnetic field is $\pm\mu_B B$ where the positive sign corresponds to the case spin angular momentum component parallel to the field. (The magnetic moment is then antiparallel!)

Let us now consider the implications on the behaviour of conduction electrons in a metal. In zero field, (Fig. 3.3(*a*)) equal numbers of electrons will occupy spin states parallel and antiparallel to the direction in which the field will be applied. (It is an important quantum mechanical concept that the spin angular momentum is always quantized in the z direction. In the absence of a magnetic field this z direction is not specified and can be chosen arbitrarily.) When a magnetic field is applied to the sample, conduction electrons with spin parallel to the field are shifted in energy with respect to those antiparallel to the field. The two distributions are displaced by $\pm\mu_B B$ with respect to the zero of energy (Fig. 3.3(*b*)). This is not a stable state as electrons in parallel spin states just above E_F can make transitions to empty antiparallel spin states just below E_F, reducing the total energy of the system. Equilibrium is re-established when the Fermi energies of the parallel and antiparallel distributions coincide (Fig. 3.3(*c*)).

The number of electrons transferred is then $\frac{1}{2}\mu_B BD(E_F)$ where $D(E_F)$ is the density of states at the Fermi energy and given by Equation (3.8). (The factor of $\frac{1}{2}$ arises because half the electrons are in parallel states and half are in antiparallel states.) As there are now $\mu_B BD(E_F)$ more electrons in the antiparallel spin state than in the parallel state this means that there is a net magnetic moment associated with the electron distribution. This net magnetic moment is parallel to the applied field because the electron spin angular momentum vector is antiparallel to the magnetic moment vector. The magnetization M of a material is defined as the magnetic dipole moment per unit volume, and thus we have

$$M = \mu_B \cdot \mu_B BD(E_F)/V. \tag{3.18}$$

Using Equation (3.8) we have

$$M = 3n\mu_B^2 B/2E_F \tag{3.19}$$

where $n = N/V$, the number of conduction electrons per unit volume. If the volume susceptibility χ is defined as

$$\chi = M/B \tag{3.20}$$

we have

$$\chi = 3n\mu_B^2/2E_F. \tag{3.21}$$

Two points should be noted. Firstly that the magnetic moment induced is proportional to the flux density B and is in the same direction as the applied field. This phenomenon which occurs both in metals and metal salts is known as *paramagnetism*. The effect is weak in metals and if we evaluate Equation (3.21) we find that the Pauli paramagnetic susceptibility, as the effect is known, is only typically

10^{-5} MKS units m^{-3}. Secondly we note that the susceptibility is almost independent of temperature, for although $D(E_F)$ does change slightly with temperature, this is a second order effect. The result contrasts strongly with the equivalent result using classical statistics and localized electrons, where a T^{-1} dependence is predicted, and indeed observed, in the insulating salts of many metals.

3.4 Chemical properties

3.4.1 Photoemission

The energy difference between the vacuum potential, at which an electron just escapes from the metal surface and the Fermi level in a metal is known as the work function, usually denoted ϕ. It is possible to measure the work function by studying the process of photoemission, that is the emission of electrons from the surface of a metal under irradiation with light. The free electron model is not too satisfactory in treating this phenomenon, as a perfectly free electron cannot totally absorb a photon due to the inability to conserve both energy and momentum in the process. However, if we accept that other collective effects make this absorption possible, we can use the model to predict the threshold frequency for photoemission. From Fig. 3.4 we can readily see that for $h\nu < \phi$, no electrons will be released from the surface as insufficient energy is available to excite electrons at the Fermi level to the vacuum level. However for $h\nu > \phi$, electrons will be excited out of the potential well of the solid and will be emitted from the surface of the metal. The maximum kinetic energy E_{kin} of the ejected photoelectrons (corresponding to electrons at the Fermi level)

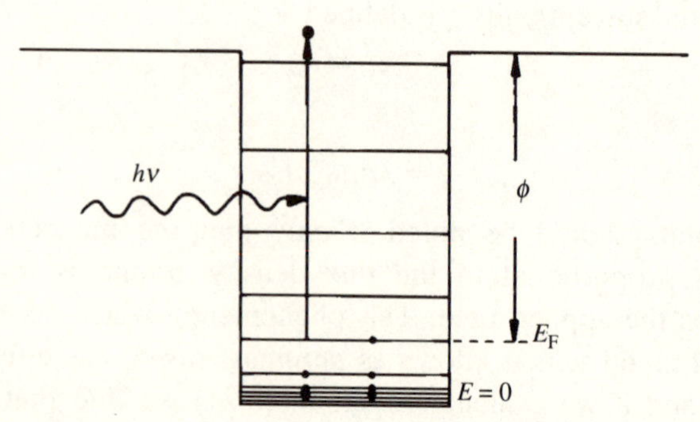

Fig. 3.4 Schematic representation of the principle of photoemission.

is accordingly

$$E_{\mathrm{kin}} = h\nu - \phi. \qquad (3.22)$$

Measurement of the threshold frequency provides a measure of ϕ. While the above discussion implied $T = 0$, there is actually a smearing out of the threshold frequency due to thermal excitation. This does not change the cut-off frequency significantly for most measurements as $\phi \gg k_B T$ at room temperature.

3.4.2 Thermionic emission

The phenomenon of thermionic emission is somewhat related to photo-emission. As a result of thermal excitation, a very small fraction of the electrons in the tail of the Fermi–Dirac distribution will have sufficient energy to escape from the potential well of the solid. This fraction will be negligibly small at room temperature, but at 900 °C, the fraction is significant. If we take the Fermi energy to be 4 eV and the work function to be 1 eV, we see that the probability of an electron having enough energy to escape (i.e. 5 eV), given by $\{\exp[(E - E_F)/k_B T] + 1\}^{-1}$, is about 4×10^{-18} at room temperature. However at 900 °C this rises dramatically to 4×10^{-5}. At these temperatures therefore a substantial number of electrons can be 'boiled off' the surface of the metal and this phenomenon has seen extensive use, for example in television tubes.

In order to determine the variation of the thermionic current with temperature, we examine the flux of electrons escaping from a surface of the solid parallel to the (yz) plane. If we define, as before, the zero of kinetic energy as that of a free electron at rest at the bottom of the Fermi sea, we can see immediately that for an electron to escape it must have an x component of velocity v_x such that

$$\tfrac{1}{2}mv_x^2 > E_F + \phi. \qquad (3.23)$$

For an electron that *just* manages to escape, i.e. has just enough kinetic energy to balance the potential energy, we can write

$$\tfrac{1}{2}mv_{x0}^2 = E_F + \phi, \qquad (3.24)$$

where v_{x0} is the x component of velocity of such an electron. The number of electrons with x component velocities between v_x and $v_x + \mathrm{d}v_x$ crossing a plane of unit area normal to the x axis per unit time is $v_x n(v_x)\mathrm{d}v_x$, where $n(v_x)\mathrm{d}v_x$ is the number of electrons with x component velocities between v_x and $v_x + \mathrm{d}v_x$. The thermionic current density flowing across the surface is then $ev_x n(v_x)\mathrm{d}v_x$. Thus the total emission current density J is given by

$$J = \int_{v_{x0}}^{\infty} ev_x n(v_x)\mathrm{d}v_x. \quad (3.25)$$

We now have to determine $n(v_x)$. This can be done straightforwardly if we refer back to Section 2.1.4. There, following Equations (2.32) we noted that there is one allowed energy level in volume $8(\pi/L)^3$ of wavevector (k) space. When the two possible spin states are included, this gives one allowed quantum state per $2(2\pi/L)^3$ of k space. Thus the number of allowed quantum states in a cube of wavevector space between k_x and $k_x + \mathrm{d}k_x$, k_y and $k_y + \mathrm{d}k_y$, k_z and $k_z + \mathrm{d}k_z$ is $2(2\pi/L)^3\mathrm{d}k_x\mathrm{d}k_y\mathrm{d}k_z$. The number of electrons per unit volume with wavevectors in this range is then

$$n(k_x, k_y, k_z)\mathrm{d}k_x\mathrm{d}k_y\mathrm{d}k_z = \frac{2(2\pi)^3\mathrm{d}k_x\mathrm{d}k_y\mathrm{d}k_z}{\exp[(E - E_\mathrm{F})/k_\mathrm{B}T] + 1}. \quad (3.26)$$

Now $m^*v_x = \hbar k_x$, $m^*v_y = \hbar k_y$ and $m^*v_z = \hbar k_z$ and therefore the number of electrons with velocity between v_x and $v_x + \mathrm{d}v_x$, v_y and $v_y + \mathrm{d}v_y$, v_z and $v_z + \mathrm{d}v_z$ is $n(v_x, v_y, v_z)\mathrm{d}v_x\mathrm{d}v_y\mathrm{d}v_z = n(k_x, k_y, k_z)$ $\mathrm{d}k_x\mathrm{d}k_y\mathrm{d}k_z$. Therefore

$$n(v_x, v_y, v_z)\mathrm{d}v_x\mathrm{d}v_y\mathrm{d}v_z = 2(2\pi)^3\left(\frac{m^*}{\hbar}\right)^3\frac{\mathrm{d}v_x\mathrm{d}v_y\mathrm{d}v_z}{\exp[(E - E_\mathrm{F})/k_\mathrm{B}T] + 1}. \quad (3.27)$$

The number of electrons with velocity between v_x and $v_x + \mathrm{d}v_x$ is then given by the integral of $n(v_x, v_y, v_z)\mathrm{d}v_x\mathrm{d}v_y\mathrm{d}v_z$ over v_y and v_z. Explicitly,

$$n(v_x)\mathrm{d}v_x = 2\left(\frac{m^*}{h}\right)^3\mathrm{d}v_x\int_{\infty}^{-\infty}\int_{-\infty}^{\infty}\frac{\mathrm{d}v_y\mathrm{d}v_z}{\exp[(E - E_\mathrm{F})/k_\mathrm{B}T] + 1}. \quad (3.28)$$

The work function ϕ for metals is much greater than $k_\mathrm{B}T$ and thus the Fermi–Dirac distribution function can be approximated to $\exp[(E_\mathrm{F} - E)/k_\mathrm{B}T]$, i.e. we can neglect the 1 with respect to the exponential. As $E = \frac{1}{2}m^*(v_x^2 + v_y^2 + v_z^2)$, Equation (3.28) becomes

$$n(v_x)\mathrm{d}v_x = 2\left(\frac{m^*}{h}\right)^3\exp(E_\mathrm{F}/k_\mathrm{B}T)\exp(-\tfrac{1}{2}m^*v_x^2/k_\mathrm{B}T)\mathrm{d}v_x \times\int_{-\infty}^{\infty}\exp(-\tfrac{1}{2}m^*v_y^2/k_\mathrm{B}T)\mathrm{d}v_y\int_{-\infty}^{\infty}\exp(-\tfrac{1}{2}m^*v_z^2/k_\mathrm{B}T)\mathrm{d}v_z. \quad (3.29)$$

On evaluation we obtain

$$n(v_x)\mathrm{d}v_x = 2\left(\frac{m^*}{h}\right)^3\left(\frac{2\pi k_\mathrm{B}T}{m^*}\right)\exp(E_\mathrm{F}/k_\mathrm{B}T)\exp(-\tfrac{1}{2}mv_x^2/k_\mathrm{B}T)\mathrm{d}v_x. \quad (3.30)$$

Therefore the current density J is given by

$$J = (4\pi em^{*2}k_{\mathrm{B}}T)/h^3)\exp(E_{\mathrm{F}}/k_{\mathrm{B}}T)\int_{v_{x0}}^{\infty}\exp(-\tfrac{1}{2}mv_x^2/k_{\mathrm{B}}T)\mathrm{d}v_x$$
$$= (4\pi em^*k_{\mathrm{B}}^2T^2/h^3)\exp(E_{\mathrm{F}}/k_{\mathrm{B}}T)\exp(-\tfrac{1}{2}mv_{x0}^2/k_{\mathrm{B}}T)$$
$$= (4\pi em^*k_{\mathrm{B}}^2T^2/h^3)\exp(-\phi/k_{\mathrm{B}}T) \quad (3.31)$$

using Equation (3.24).

If we write $A = 4\pi em^*k_{\mathrm{B}}^2/h^3$ we have

$$J = AT^2\exp(-\phi/k_{\mathrm{B}}T) \quad (3.32)$$

which is known as the Dushman–Richardson equation. This relation provides a means of determining the work function ϕ from a plot of $\ln(J/T^2)$ versus $1/T$. However, great care must be taken to work under clean conditions if a meaningful value of ϕ is to be obtained. Equation (3.32) shows quantitatively how the thermionic emission current rises dramatically with temperature, the exponential term dominating the temperature dependence.

3.4.3 Contact potential

The work functions of metals will generally differ, just as the Fermi energies differ. Therefore if two pieces of dissimilar metal are placed in contact there is initially a difference between the Fermi levels on either side of the interface (Fig. 3.5(*a*)). This is not an equilibrium situation and an electron flow takes place from metal 1 with a lower work function ϕ_1 to metal 2 with the higher work function ϕ_2 in order to minimize the total energy of the system. This energy flow can take

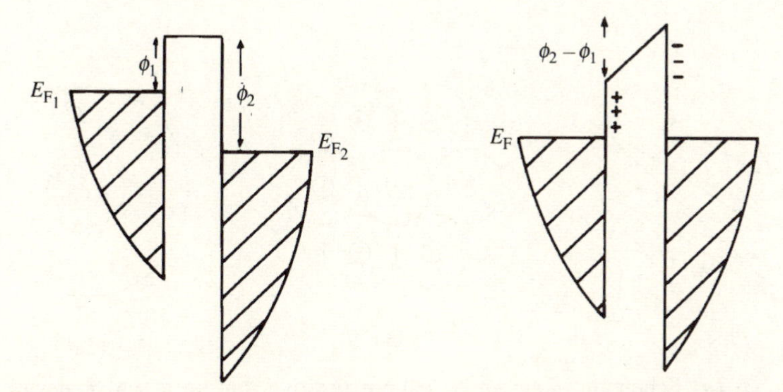

Fig. 3.5 (*a*) Non-equilibrium distribution when two dissimilar metals are brought in contact. (*b*) Equilibrium distribution showing dipole layer at the interface.

place either by quantum mechanical tunnelling through the interface of thermionic excitation. However, this transfer of charge leaves metal 1 positively charged while metal 2 becomes negatively charged. An electric dipole layer is created across the interface layer, changing the relative potential between the two metals. The electrostatic potential in the interface region is then no longer constant. Electron flow continues until the potential difference associated with the dipole layer is sufficient to bring the Fermi levels on both sides of the interface into coincidence. The potential difference across the interface, called the contact potential, V can be seen from Fig. 3.5(b) to be $(\phi_2 - \phi_1)/e$. We note that it depends only on the difference in contact potentials, not on the depth of the energy wells.

3.4.4 Cohesive energy

We are now in a position to examine the nature of the forces which hold a metal together. It is readily seen that our model free electron metal is electrically neutral, but is it the minimum energy configuration? In order to answer this question we must first examine what electrical forces are involved. To do this we consider an idealized metal called *jellium*. This model system consists of a lattice of positive ions surrounded by a uniform distribution of free conduction electrons (Fig. 3.6) which preserve electrical neutrality. To a very good approximation, the conduction electrons screen the positive charges and the repulsive forces between neighbouring ions can be neglected. An attractive force arises from the interaction of the positive ions with the surrounding sea of negative charge and we can model this by calculat-

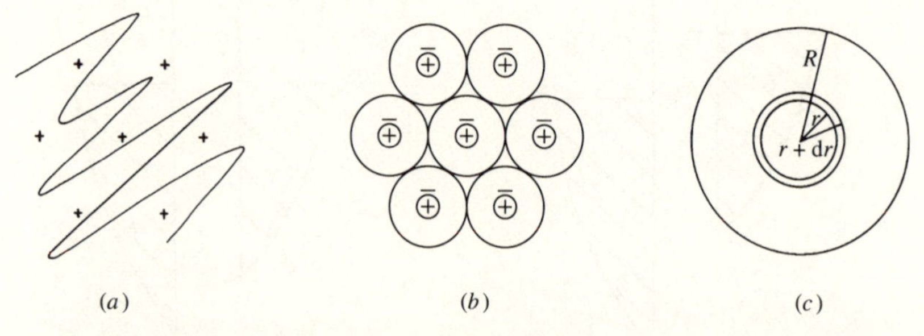

Fig. 3.6 (a) Schematic diagram of jellium positive ions in a sea of negative charge. (b) Positive ions surrounded by spheres of negative charge, radius R. (c) Concentric shell between radius r and $r + \mathrm{d}r$.

ing the interaction energy between a point positive charge $+e$ and a surrounding sphere of negative charge $-e$. The volume of the sphere is taken to be the volume occupied by one atom. A repulsive force arises from the kinetic energy of the conduction electrons. Equations (2.37) and (2.38) can be combined to give the average energy of a three-dimensional free electron gas, namely,

$$\langle E \rangle = \frac{3\hbar^2}{10m^*}(3\pi^2 N/V)^{2/3}, \tag{3.33}$$

where N is the total number of electrons and V is the sample volume. As the electron concentration per unit volume (N/V) increases, so does the average kinetic energy. This gives rise to a repulsive force tending to spread out the negative charge. The total energy of the system is the sum of the attractive and repulsive energies and its minimum represents the stable state.

We can calculate the attractive energy straightforwardly. Let the radius of the sphere of negative charge $-e$ be R, and consider the energy associated with a shell of charge, thickness $\mathrm{d}r$ between radius r and $r + \mathrm{d}r$. The charge inside the shell $q(r)$ is

$$q(r) = e - e(r^3/R^3) \tag{3.34}$$

and the electrostatic potential at the shell is $q(r)/4\pi\varepsilon_0 r$, ε_0 being the permitivity of free space. Therefore the contribution to the attractive energy $\mathrm{d}U_\mathrm{A}$ is given by

$$\mathrm{d}U_\mathrm{A} = 4\pi r^2 \rho \mathrm{d}r[e - e(r/R)^3]/4\pi\varepsilon_0 r \tag{3.35}$$

where the negative charge density ρ is given by

$$-e = 4\pi\rho R^3/3. \tag{3.36}$$

The total energy associated with the Coulomb attraction U_A is thus

$$\begin{aligned} U_\mathrm{A} &= -\frac{4\pi 3e^2}{4\pi R^3 4\pi\varepsilon_0}\int_0^R [r - (r^4/R^3)]\mathrm{d}r \\ &= -9e^2/40\pi\varepsilon_0 R. \end{aligned} \tag{3.37}$$

Equation (3.33) gives the average kinetic energy of the free electron gas, and noting that there is charge equivalent to one electron in volume $4\pi R^3/3$ we have the repulsive energy U_R given by

$$\begin{aligned} U_\mathrm{R} &= \frac{3\hbar^2}{10m^*}\left(\frac{3\pi^2 3}{4\pi R^3}\right)^{2/3} \\ &= \frac{3\hbar^2}{10m^* R^2}\left(\frac{9\pi}{4}\right)^{2/3}. \end{aligned} \tag{3.38}$$

Table 3.1

	Jellium	Gold	Copper	Nickel
Structure	face centred cubic	face centred cubic	face centred cubic	face centred cubic
Lattice parameter (Å)	3.68	4.07	3.61	3.52
R_0 (Å)	1.30	1.44	1.28	1.25

The sum of U_A and U_R gives the total energy per atom volume for the solid, that is,

$$U = U_A + U_R = \frac{3\hbar^2}{10m^*R^2}\left(\frac{9\pi}{4}\right)^{2/3} - \frac{9e^2}{40\pi\varepsilon_0 R}. \tag{3.39}$$

For equilibrium U must be a minimum with respect to R, and this permits derivation of an equilibrium value R_0. At the minimum

$$\frac{\partial U}{\partial R} = 0 = \frac{-6\hbar^2}{10m^*R^3}\left(\frac{9\pi}{4}\right)^{2/3} + \frac{9e^2}{40\pi\varepsilon_0 R^2} \tag{3.40}$$

and this yields

$$R_0 = \frac{2\hbar^2 4\pi\varepsilon_0}{3m^*e^2}\left(\frac{9\pi}{4}\right)^{2/3}. \tag{3.41}$$

Evaluation of Equation (3.41) gives

$$R_0 = 2.45R_h = 1.30 \text{ Å}.$$

where $R_h = \hbar^2 4\pi\varepsilon_0/e^2m^*$, the Bohr radius of the hydrogen atom in the ground state. This is a remarkably good estimate for the spacings of atoms in metal crystals. Values of R_0 deduced from the crystal lattice parameters of some metals are given in Table 3.1.

Equation (3.41) can be used in Equation (3.39) to give the total equilibrium energy U_0.

We have

$$U_0 = -9e^2/80\pi\varepsilon_0 R_0. \tag{3.42}$$

On evaluation we find $U_0 \approx -5$ eV per atom, a value very similar to that found experimentally. The energy U_0 is negative, explaining why it is that metals hold together. Because it represents the energy required to separate the solid into its constituent atoms it is known as the cohesive energy.

3.5 Mechanical properties

3.5.1 Compressibility of jellium

When the electron concentration is increased, the average energy of an electron in the conduction electron gas rises. From Equation (3.38) we note that the total kinetic energy of the free electron gas rises as R^{-2}. This has a faster rate of fall than the Coulomb energy associated with the attraction between positive ions and electrons, which varies as R^{-1}. The two energy terms are plotted in Fig. 3.7. For an atomic radius larger than R_0 the attractive term dominates, below R_0 the repulsive term dominates. Therefore, if we compress our sample of jellium, although the Coulomb attractive force will be increased, the repulsive force arising from the increased kinetic energy of the

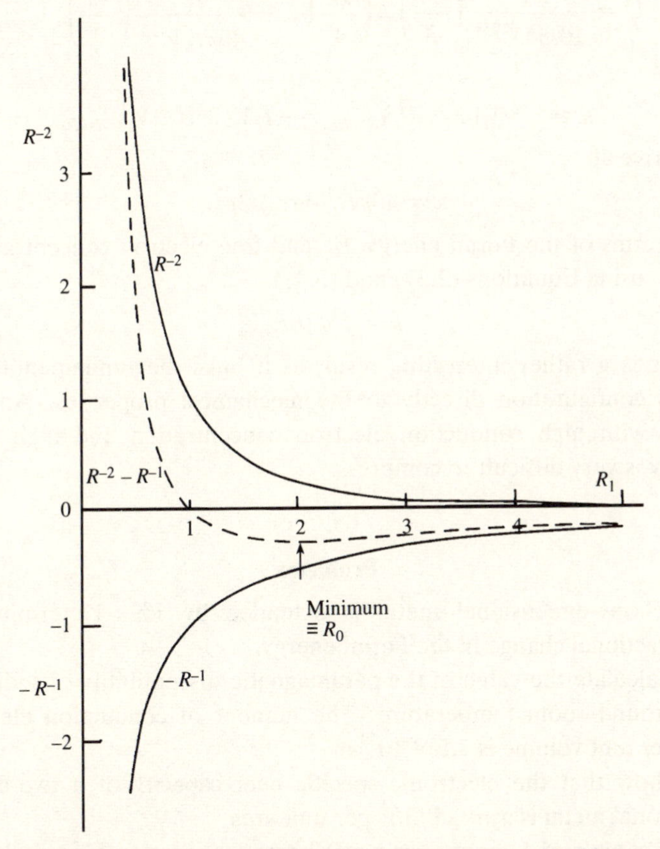

Fig. 3.7 Function form of repulsive (R^{-2}) and attractive (R^{-1}) potential of jellium. On addition, a shallow minimum is found at R_0.

conduction electrons will increase more rapidly, giving a resistance to compression.

Thermodynamically we define this resistance to compression in terms of the compressibility κ where

$$\kappa = -\partial(\ln V)/\partial P, \tag{3.43}$$

which indicates that κ represents the fractional change in volume V for a unit change in pressure P. Now the work done on a sample $\mathrm{d}U$ when its volume is changed $\mathrm{d}V$ is given by

$$\mathrm{d}U = P\mathrm{d}V. \tag{3.44}$$

Therefore using Equations (3.39), (3.43) and (3.44) we can derive an expression for the compressibility. In terms of volume, Equation (3.39) can be written

$$U = \frac{3\hbar^2}{10m^* V^{2/3}}\left(\frac{4\pi}{3}\right)^{2/3}\left(\frac{9\pi}{4}\right)^{2/3} - \frac{9e^2}{40\pi\varepsilon_0 V^{1/3}}\left(\frac{4\pi}{3}\right)^{1/3}, \tag{3.45}$$

and as

$$\kappa = -1/V(\partial V/\partial P)_{R=R_0} = -1/V(\partial^2 U/\partial V^2)_{R=R_0} \tag{3.46}$$

we arrive at

$$\kappa = 40\pi R_0^4(4\pi\varepsilon_0)/3e^2. \tag{3.47}$$

In terms of the Fermi energy E_F and free electron concentration n this is, using Equations (2.37) and (3.41)

$$\kappa = 15/2nE_F. \tag{3.48}$$

This is a rather interesting result as it links the fundamental electronic configuration directly to the mechanical properties. An ideal metal with high conduction electron concentration and high Fermi energy is very difficult to compress.

Problems

3.1 A one-dimensional metal is extended by 1%. Determine the fractional change in the Fermi energy.

3.2 Calculate the value of the paramagnetic susceptibility of sodium at around room temperature. The number of conduction electrons per unit volume is 2.5×10^{28} m^{-3}.

3.3 Show that the electronic specific heat capacity of a two-dimensional metal is $\pi m k_B^2 T/3\hbar^2$ per unit area.

3.4 A sample of a copper has a small amount of iron added so that the iron atoms are individually randomly dispersed through the material. If the free electron concentration in copper is $8.45 \times 10^{28}\ \mathrm{m}^{-3}$, the Fermi energy is 7 eV and the residual resistivity at very low temperature is 3×10^{-7} ohm m, determine the concentration of iron in the sample. (You may assume that the residual resistivity of pure copper is zero.)

3.5 Prove that the compressibility $-(\partial V/\partial P)V^{-1}$ of an ideal three-dimensional Fermi gas is $3/2nE_{\mathrm{F}}$ where n is the conduction electron concentration. (Note the difference between this and Equation (3.48), where the effect of the ion cores is included.)

4
Energy bands

4.1 The Hall effect

In Chapter 3 we saw that the quantum mechanical free electron theory gave a remarkably good explanation of a number of properties of simple metals. Considering the simplicity of the model, this is a notable achievement. However there are some experimental phenomena which the free electron theory is unable to explain. One of these is the Hall effect.

The Hall effect is now included in many secondary school level syllabuses and is a specific case of a number of magnetoresistive effects. Fig. 4.1(*a*) shows the experimental geometry and coordinate system used in this discussion. The specimen is cut as a long rod with long faces perpendicular to the y and z directions. A current of density J_x is passed along the rod, placed in a magnetic field of flux density B_y. Due to the Lorenz force acting on the current carriers, they are deflected towards one of the surfaces of the specimen (Fig. 4.1(*b*)). If the current carriers are electrons they are deflected as shown. (We use the left hand rule and note that conventional and electron currents are

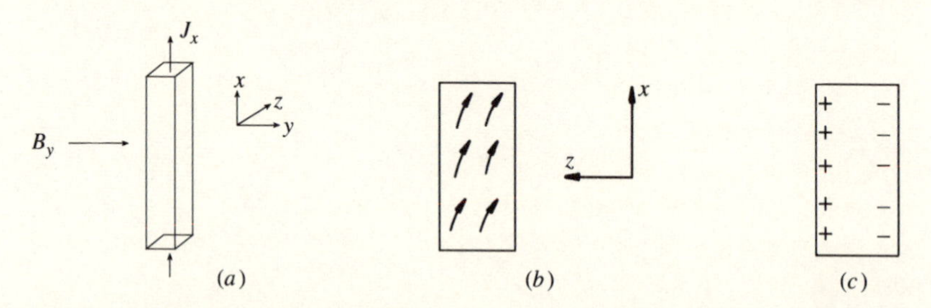

Fig. 4.1 The Hall effect. (*a*) Schematic diagram showing the experimental geometry. (*b*) Deflection of electrons due to the Lorenz force. (*c*) Change in charge density giving rise to electric field within sample.

in opposite senses.) This deflection causes a build up of negative charge on the $-z$ face, leaving a deficiency of charge on the $+z$ face (Fig. 4.1(*c*)). Because of this redistribution of charge density, an electric field E_z is set up. In turn this electric field produces a force on the current carriers in the opposite sense to the Lorenz force. In equilibrium these electric and magnetic forces are equal. The total force on a current carrier of charge e and velocity $\mathbf{v}$ is thus

$$\mathbf{F} = e(\mathbf{E} - \mathbf{v} \times \mathbf{B}) \tag{4.1}$$

or

$$F_z = e(E_z - v_x B_y) \tag{4.2}$$

where v_x is the x component of velocity. Now the current density J_x is given by

$$J_x = nev_x \tag{4.3}$$

and so

$$F_z = e(E_z - J_x B_y/ne). \tag{4.4}$$

As $F_z = 0$ in equilibrium, this gives

$$E_z/J_x B_y = 1/ne = R_{\mathrm{H}} \tag{4.5}$$

where R_{H} is known as the Hall coefficient. It can be measured very easily because the redistribution of charge density resulting in an electric field E_z leads to a potential difference appearing across the $\pm z$ faces of the sample. From this Hall voltage, the flux density and the current density we can deduce both the carrier concentration and the sign of the charge on the current carriers. We expect R_{H} to be negative, as the electron charge is negative, and indeed for many metals such as copper, silver and sodium this is what is found (Table 4.1). However, it is impossible to reconcile the above discussion with the experimentally measured Hall coefficients of aluminium and indium. Like the Hall coefficients of a number of other metals, these are positive, suggesting that the current carriers have POSITIVE charge. Our present model is incapable of resolving this discrepancy. It appears that there is one current carrier per atom but it is of the wrong sign. Measurement of e/m of the current carriers in aluminium confirms that the mass is almost that of the free electron. The resolution of this paradox lies in the appreciation of the effects of the periodic potential introduced by the crystal structure. That is, we must no longer consider the electron to be free in a flat bottomed potential but must consider the potential to be perturbed in a periodic manner. We must go one step further in complexity in the model.

Table 4.1

Element	Experimental R_H ($\times 10^{-10}$ m^3/C)	Calculated R_H ($\times 10^{-10}$ m^3/C)	Assumed no. of electrons per atom
Na	−2.36	−2.34	1
Cu	−0.54	−0.74	1
Ag	−0.9	−1.07	1
In	+1.60	−1.60	1
Al	+1.02	−1.04	1

4.2 Crystal structure

While it had long been guessed that crystals were composed of regular arrays of atoms, it was not until Friedrich and Knipping, working under von Laue, produced the first X-ray diffraction patterns in 1912 that the foundations of modern crystallography were laid. From the seventeenth century onwards, the beautiful symmetry of natural crystals had led mineralogists to speculate on the size and nature of the building blocks from which crystals were formed. When Friedrich and Knipping obtained their X-ray diffraction patterns it became clear that the spacing between atoms in crystals was of the order of the X-ray wavelengths, namely about 10^{-10} m.

It was found that the simple metals, for example copper, aluminium and silver consisted of a close packed array of atoms. There are no large spaces between atoms, and the atoms pack in such a way as to minimize the total energy of the array. We have seen in our treatment of jellium (in Chapter 3) how attractive and repulsive forces balance to give an equilibrium atomic volume. In this light, it is surprising that the free electron model works so well. However, in the free electron model we assume only one free conduction electron per atom. This assumes that the valence electrons of the atoms become the conduction electrons and therefore there will be many core electrons in lower energy levels bound to the ions. These localized core electrons screen the electrostatic potential of the ion and it does not act over a long range. Despite close packing of atoms, the potential well can then be adequately represented by a flat bottomed potential well with narrow spikes at the ion cores. To a first approximation we can treat this periodic potential as a perturbation of the free electron model potential.

Probably the most widely known effect, apart from the symmetries of crystal faces, arising from the periodic nature of crystal structure is the diffraction of X-rays and electrons. It is important to stress that almost all metallic materials in common use are crystalline, although not in single crystal form. For example, engineering steels, used for manufacturing a host of products from pins to motor cars, consist of small crystalline grains, typically a few micrometres in diameter. As the crystal axes in neighbouring grains are not aligned, the net result is an almost random distribution of crystal orientations over the whole specimen. Nevertheless, the material is crystalline and with the correct geometry, X-ray diffraction patterns can be recorded.

We will now go on to include the effects of crystal structure in our model. However, before this can be done, we must describe what we mean by crystal structure. In such a description it is important to separate out two distinct concepts which underpin the whole of crystallography. These are:

1 The *lattice*, consisting of points in a geometrical array
2 The *basis*, which describes the arrangement of atoms associated with each lattice point.

4.2.1 The lattice vectors

These are the fundamental vectors which define the lattice such that each point in the lattice is described by a lattice translation vector. Let us take a two-dimensional example, shown in Fig. 4.2. Let **a** and **b** be vectors from the origin 0 such that the tips of the vectors define lattice points A and B. When another translation of **a** is made from A we reach another lattice point A' etc. We can construct a two-dimensional lattice from these two unit vectors. In three dimensions three vectors, **a**, **b** and **c**, are needed to create such a lattice.

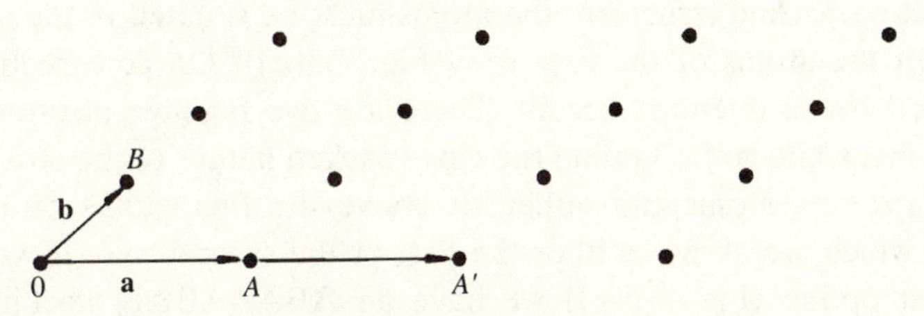

Fig. 4.2 A two-dimensional lattice created by the translation vectors **a** and **b**.

Thus we construct a lattice of *points* which can be described by its symmetry. For example, a simple cubic lattice has a fourfold rotation symmetry about a cube edge direction and threefold symmetry about the body diagonal. It also has mirror symmetry about an axis normal to a cube face. In three dimensions we can find 14 different types of lattice, called *Bravais lattices*. For the present purposes we will restrict ourselves principally to cubic systems as, fortunately, all of the important semiconductor materials are cubic.

4.2.2 Basis vectors

These describe the position of atoms associated with each lattice point. For example if we have one atom at each simple cubic lattice point then the basis vector would be [000]. However, let us imagine a face centred cubic lattice, where there are lattice points in the cube faces. Now we could again have only one atom at each site, and that would be a face centred cubic structure like copper. If on the other hand there were two atoms associated with each lattice point, we could take a basis of two atoms with basis vectors [000] and $[\frac{1}{4}\frac{1}{4}\frac{1}{4}]$, to create a structure called the diamond cubic structure, of which silicon has to be the most important member. Note that there may be more than one possible set of lattice and basis vectors for a given structure. With so many possibilities we will consider only the four simplest structures.

Simple cubic

The simple cubic structure is not close packed and is thus extremely rare. Only *polonium* has a simple cubic structure.

Hexagonal close packed

If we pack a set of spheres as closely as possible in a plane, we create a hexagonal array (Fig. 4.3(*a*)). The lattice translation vectors are easy to see and are at 60° to each other. Now let us construct a second row. For a close packed structure, the atoms must be situated in the spaces between the atoms of the first row (Fig. 4.3(*b*)). On construction of the third row a dilemma occurs. There are two possible positions for the atoms, while still retaining the close packed nature of the structure. The third row atoms can either sit above the first atoms or in the spaces which are above neither the first or the second row. If we take the first option (Fig. 4.3(*c*)) we have an ABABABAB stacking sequence. This is the hexagonal close packed (HCP) structure.

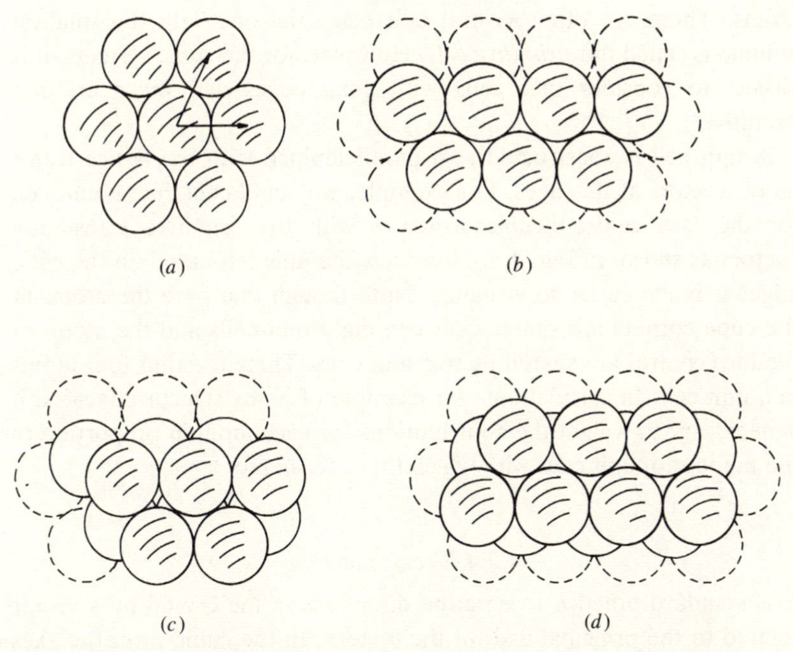

Fig. 4.3 (*a*) A two-dimensional close packed array. (*b*) Array of spheres located above interstices in the first row. (*c*) The hexagonal close packed structure. (*d*) The face centred cubic structure.

Face centred cubic

If, in constructing the third row, we had chosen the second option where the atoms are sited in the spaces which do not correspond to the atom positions of the first row, we would have had an ABCABCABC sequence (Fig. 4.3(*d*)). This forms the face centred cubic (FCC) structure. Of course, because we have hexagonal symmetry, we must be looking at the planes perpendicular to the body diagonal.

Body centred cubic

This is not a close packed structure, but is nearly so. Iron is the most important material with this (BCC) structure.

4.2.3 Unit cells

Crystalline periodicity can be described in terms of *units cells* which are the fundamental repeat blocks of the structure. The crystal is composed of many unit cells, all of which contain the same distribution of atoms. There are often several unit cells, the one with the smallest volume is called the *primitive cell*. However, for the cubic systems, it is easiest to consider unit cells with cube edges and these are not primitive.

A unit cell is constructed as a parallelepiped with the lattice translation vectors as its edges. For example, we could construct a unit cell for the face centred cubic structure with the primitive translation vectors as shown in Fig. 4.4. However, the unit cell based on the cube edges is much easier to visualize. Note though that here the atoms at the cube corners are shared between eight unit cells and the atoms at the face centres are shared by two unit cells. There are thus four atoms in a unit cell. In calculations, for example of X-ray structure factors, it is necessary to weight the contributions for each atom in proportion to the number of unit cells with which they are shared.

4.2.4 Vector notation

It is standard practice to describe directions in the crystal by a vector related to the principal axes of the system. In the cubic case, the axes are orthogonal and parallel to the cube edges. By convention, a vector is denoted by indices contained in square brackets. Thus, $[u\,v\,w]$ is a vector $u\mathbf{a}_1$, $v\mathbf{a}_2$, $w\mathbf{a}_3$ in the cubic axis system $\mathbf{a}_1$, $\mathbf{a}_2$, $\mathbf{a}_3$. Use of square brackets denotes a specific vector. When a negative number, e.g. $-u$, is required for the component vector $-u\mathbf{a}_1$, then it is customary to

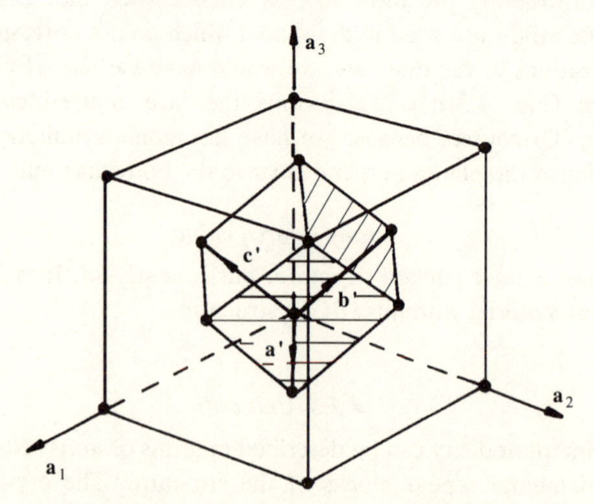

Fig. 4.4 Primitive cell of a face centred cubic lattice.

place a *bar* over the u. Thus $[\bar{u}\,v\,w]$ corresponds to $-u\mathbf{a}_1$, $v\mathbf{a}_2$, $w\mathbf{a}_3$. A general vector of the form $\pm u\mathbf{a}_1$, $\pm v\mathbf{a}_2$, $\pm w\mathbf{a}_3$ is denoted with angle brackets $\langle \mathrm{u\,v\,w} \rangle$. For example $\langle 1\,1\,1 \rangle$ denotes any body-diagonal vector in the cubic system which could be $[1\,1\,1]$ or $[1\,\bar{1}\,1]$ or $[1\,1\,\bar{1}]$ etc. (see the second worked example in this chapter).

4.2.5 Miller indices

Planes of atoms in crystals are most important in determining the anisotropic physical properties of crystals, such as plastic deformation. A plane is characterized by the intersections of the plane with the principal axes. Suppose the intersection is at $u\mathbf{a}_1$, $v\mathbf{a}_2$, $w\mathbf{a}_3$ (Fig. 4.5). The inverse values of the intersections are $1/u$, $1/v$, $1/w$ and it is then practice to reduce these to the simplest integer set h, k, l. The plane is then labelled by these Miller indices $(h\,k\,l)$. By convention, round brackets are used for planes and again a bar over an integer denotes a minus sign. In a similar manner to that for vectors, a family of planes with Miller indices $\pm h$, $\pm k$, $\pm l$ is denoted $\{h\,k\,l\}$. In summary, a

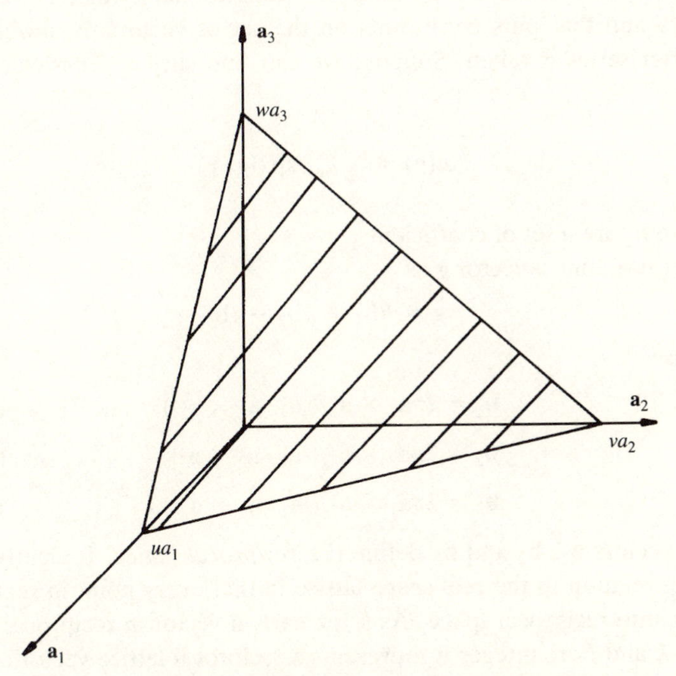

Fig. 4.5 Basis of Miller index notation for planes.

plane of Miller indices $(h\,k\,l)$ cuts the crystal axes at $1/h$ along the $\mathbf{a}_1$ axis, $1/k$, along the $\mathbf{a}_2$ axis and $1/l$ along the $\mathbf{a}_3$ axis.

In the cubic system a vector $[h\,k\,l]$ is perpendicular to the plane of Miller indices $(h\,k\,l)$. It is also straightforward to show geometrically that the spacing d between planes of Miller indices $\{h\,k\,l\}$ in the cubic system is given by

$$d = a/(h^2 + k^2 + l^2)^{1/2} \tag{4.6}$$

where a is the length of the edge of the cubic unit cell.

4.2.6 The reciprocal lattice construction

The important thing to remember at all times about a crystal is that the electron density $n(\mathbf{r})$ at a point defined by the vector $\mathbf{r}$ is exactly the same as that at a point $\mathbf{r} + \mathbf{T}$ where $\mathbf{T}$ is a lattice translation vector. Thus

$$n(\mathbf{r}) = n(\mathbf{r} + \mathbf{T}). \tag{4.7}$$

Now because the medium is periodic, it is possible to expand the electron density $n(\mathbf{r})$ as a series called a Fourier series over a set of vectors. This series must retain the translational symmetry required above and that puts constraints on the set of vectors over which the Fourier series is taken. Suppose we can find such a set of vectors $\mathbf{g}$, then

$$n(\mathbf{r}) = \sum_g n_g \exp(\mathrm{i}\mathbf{g}\cdot\mathbf{r}) \tag{4.8}$$

where n_g are a set of coefficients.

Let us define a vector $\mathbf{g}$ as

$$\mathbf{g} = h\mathbf{b}_1 + k\mathbf{b}_2 + l\mathbf{b}_3 \tag{4.9}$$

where

$$\mathbf{b}_1 = 2\pi\mathbf{a}_2 \times \mathbf{a}_3/(|\mathbf{a}_1 \cdot \mathbf{a}_2 \times \mathbf{a}_3|) \tag{4.10a}$$

$$\mathbf{b}_2 = 2\pi\mathbf{a}_3 \times \mathbf{a}_1/(|\mathbf{a}_1 \cdot \mathbf{a}_2 \times \mathbf{a}_3|) \tag{4.10b}$$

$$\mathbf{b}_3 = 2\pi\mathbf{a}_1 \times \mathbf{a}_2/(|\mathbf{a}_1 \cdot \mathbf{a}_2 \times \mathbf{a}_3|). \tag{4.10c}$$

The vectors $\mathbf{b}_1$, $\mathbf{b}_2$ and $\mathbf{b}_3$ define the *reciprocal lattice*. It clearly has a direct relation to the real space lattice in that every point in real space maps into reciprocal space. As $\mathbf{g}$ is clearly a vector in reciprocal space, if h, k and l are integer it represents a reciprocal lattice vector and has a direct relation to the crystal lattice. (Note that reciprocal lattice vectors have dimensions $(\text{length})^{-1}$. This is just the same as wavevectors and we will see that they are closely related.)

Let us now see why the reciprocal lattice vectors are so important. If a lattice translation vector $\mathbf{T}$ is written

$$\mathbf{T} = u\mathbf{a}_1 + v\mathbf{a}_2 + w\mathbf{a}_3 \tag{4.11}$$

then, if we expand the electron density at $\mathbf{r} + \mathbf{T}$, $n(\mathbf{r} + \mathbf{T})$, as a Fourier series, we have

$$n(\mathbf{r} + \mathbf{T}) = \sum_g n_g \exp(\mathrm{i}\mathbf{g} \cdot \mathbf{r}) \exp(\mathrm{i}\mathbf{g} \cdot \mathbf{T}), \tag{4.12}$$

and $\exp(\mathrm{i}\mathbf{g} \cdot \mathbf{T}) = 1$ as

$$\begin{aligned} \exp(\mathrm{i}\mathbf{g} \cdot \mathbf{T}) &= \exp[\mathrm{i}(h\mathbf{b}_1 + k\mathbf{b}_2 + l\mathbf{b}_3) \cdot (u\mathbf{a}_1 + v\mathbf{a}_2 + w\mathbf{a}_3)] \\ &= \exp[\mathrm{i}2\pi(uh + vk + wl)] \\ &= 1. \end{aligned} \tag{4.13}$$

4.2.7 Reciprocal lattice vectors and real space planes

It is easy to show that if a plane in real space has Miller indices $(h\,k\,l)$ then the reciprocal lattice vector $\mathbf{g} = h\mathbf{b}_1 + k\mathbf{b}_2 + l\mathbf{b}_3$ or (in shorthand notation) $[h\,k\,l]$ is perpendicular to it. We also find that the interplanar spacing is $2\pi/|\mathbf{g}|$. Note that a plane in real space maps into a point in reciprocal space and vice versa.

4.3 Electron scattering

4.3.1 The Laue equation

The above discussion has provided us with all the conceptual tools necessary to understand the basic physics of X-ray and electron diffraction. Let us consider the scattering of an X-ray or electron wave by a single electron. If we consider only elastic scattering, then the electron reradiates at the same frequency in all directions. Now let us calculate the scattering from an atom which consists of electrons smeared out over a volume in space. Consider an electron j in element of volume $\mathrm{d}V$ at $\mathbf{r}_j$ and let the incident beam wavevector be $\mathbf{k}_0$ and the outgoing wavevector be $\mathbf{k}'$. Then compared with an electron at the origin 0 the phase difference between a wave scattered at 0 and $\mathbf{r}_j$ is $\exp[\mathrm{i}(\mathbf{k}_0 - \mathbf{k}') \cdot \mathbf{r}_j]$. The scattered amplitude will be proportional to the total number of electrons in volume $\mathrm{d}V$ at $\mathbf{r}$ and hence we have for f, the total scattered amplitude

$$f = \int n(\mathbf{r}) \exp[\mathrm{i}(\mathbf{k}_0 - \mathbf{k}') \cdot \mathbf{r}_j] \mathrm{d}V. \tag{4.14}$$

If we write $n(\mathbf{r})$ as a Fourier series we have

$$f = \sum_g \int n_g \exp[\mathrm{i}(\mathbf{k}_0 - \mathbf{k}' + \mathbf{g}) \cdot \mathbf{r}_j] \mathrm{d}V. \tag{4.15}$$

It is easy to show that the exponential term is extremely small unless

$$\mathbf{K} = \mathbf{k}' - \mathbf{k}_0 = \mathbf{g}. \tag{4.16}$$

This is the Laue equation and we will see below that it is equivalent to the Bragg equation.

4.3.2 The Bragg equation

At the same meeting that the first experimental evidence for X-ray diffraction was announced, von Laue presented the first simple theory of X-ray diffraction. It is, however, the somewhat later theory due to Bragg which is found easier to grasp and which we examine now. The important insight which comes directly from the Bragg theory is that because the scattering of X-rays by atoms is very weak, strong diffracted beams are only observed when the scattering from successive PLANES of atoms are in phase. This is very different from the situation in a diffraction grating or indeed in X-ray diffraction from fibres, where we are considering essentially a two-dimensional scattering process. From crystals, constructive interference is required from planes of atoms and this occurs when the reflection is specular and waves scattered by successive atom planes are in phase with one another.

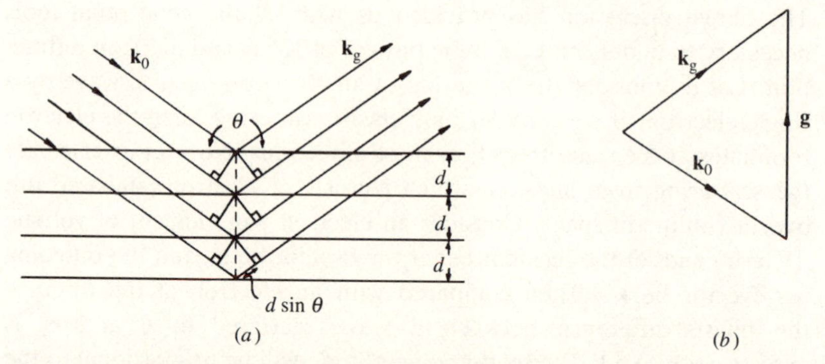

Fig. 4.6 (*a*) The origin of Bragg's Law showing constructive interference of the specular reflection from many atomic planes. (*b*) The equivalent Laue construction in reciprocal space.

As shown in Fig. 4.6(a), the path difference introduced in a plane wave specularly reflected from successive layers of atoms, is given by $2d \sin \theta$. Notice that, unlike the case of the diffraction grating, θ is defined as the angle between the incident wave propagation direction and the atom planes. When the path difference is an integral number n of the wavelength then successive scattered waves interfere constructively. This gives the Bragg equation

$$n\lambda = 2d \sin \theta \tag{4.17}$$

for a diffracted wave. For a monochromatic wave, therefore, only waves incident at certain discrete angles with respect to the crystal lattice will reinforce and give rise to a detectable diffracted beam.

Davisson and Gerner demonstrated that electrons also had a wave nature and could be diffracted. As there is nothing about the derivation of Equation (4.17) specific to X-rays, the same equation can be applied to diffraction of any radiation with wavelength less than twice the atom plane spacing. (This cut off is set by the requirement that $\sin \theta \leqslant 1$). For diffraction of electrons and neutrons, Bragg's Law (Equation (4.17)) is thus applicable. As we are concerned with the properties of electrons in solids, we shall confine our attention to the behaviour of an electron wave in a periodic solid, and more specifically to electrons whose kinetic energy is less than the potential energy associated with the solid. In other words, we do not consider the diffraction of high energy electrons which pass through the crystal in an electron microscope but the bound electrons which are responsible for the conduction processes. These too will be diffracted, but the physics is a little more subtle.

To aid the subsequent discussion let us transform Bragg's Law into the form equivalent to the Laue equation. Referring to Fig. 4.6(b), let us define wavevectors $\mathbf{k}_0$ and $\mathbf{k}_g$ for the incident and diffracted waves respectively. Because the scattering is elastic $|\mathbf{k}_0| = |\mathbf{k}_g| = k$. Let us also define a vector $\mathbf{g}$ normal to the crystal lattice planes of magnitude

$$|\mathbf{g}| = 2\pi/d. \tag{4.18}$$

As discussed previously, $\mathbf{g}$ is the reciprocal lattice vector corresponding to the planes of spacing d. Examination of the geometry of Fig. 4.6(b) shows the Laue condition (Equation (4.16)) that

$$\mathbf{k}_g - \mathbf{k}_0 = \mathbf{g} \tag{4.19}$$

where

$$|\mathbf{k}_0| \sin \theta = |\mathbf{k}_g| \sin \theta = g/2. \tag{4.20}$$

Recalling that $k = 2\pi/\lambda$, it is easy to see that Equations (4.18) and (4.20) combine to give Equation (4.17) with $n = 1$.

During the discussion of the scattering mechanisms giving rise to thermal and electrical resistivity (Section 3.2.3), we came to the conclusion that the conduction electrons were not scattered by the ion cores in the crystal. The mean free path was too long. Somehow the electrons miss the ion cores. The origin of this lies in the form of the electron wave inside the crystal. If we solve Schrödinger's equation for the case of a periodically varying potential V, then it follows that the solutions for the wavefunction ψ must also be periodic. This is a paraphrase of the Bloch theorem. These wavefunctions with a periodicity of the crystal lattice potential are known as Bloch waves. (Despite being named in the 1930s after F. Bloch, P. O. Ewald first introduced them as early as 1917. This is just one example of a lack of communication between solid state physicists and crystallographers. Beware also the definitions of k and g. Physicists define $k = 2\pi/\lambda$, crystallographers use $k = 1/\lambda$. Similarly physicists use $g = 2\pi/d$ whereas crystallographers define $g = 1/d$. Forewarned is forearmed!)

There are a number of X-ray diffraction experiments which show that X-rays really do propagate as these periodic waves in crystals. They are not just mathematical conveniences. It is therefore most important to appreciate how such a periodic wave can be formed from plane waves, and we will do this with reference to the simplest case.

Consider the situation *inside* the crystal when the Bragg condition (Equation (4.17)) is satisfied. Then we have two waves propagating, one with wavevector $\mathbf{k}_0$ and one with wavevector $\mathbf{k}_g$. Let the amplitudes be ψ_0 and ψ_g respectively. Assuming these to be plane waves, the total wave amplitude ψ, including the phase terms is then

$$\psi = \psi_0 \exp[i(\mathbf{k}_0 \cdot \mathbf{r} + \omega t)] + \psi_g \exp[i(\mathbf{k}_g \cdot \mathbf{r} + \omega t)]. \qquad (4.21)$$

$\mathbf{r}$ is a position vector and the $\exp[i(\mathbf{k} \cdot \mathbf{r})]$ term is simply a generalization to three dimensions of the $\exp[i(kx)]$ variation of a plane wave in one dimension.

Using the Laue equation (Equation (4.16)) Equation (4.21) can be rewritten

$$\psi = \{\psi_0 + \psi_g \exp[i(\mathbf{g} \cdot \mathbf{r})]\} \exp[i(\mathbf{k}_0 \cdot \mathbf{r} + \omega t)]. \qquad (4.22)$$

The total density, ρ, is given by the product of ψ with its complex conjugate ψ^*. Thus

$$\begin{aligned} \rho &= \psi^* \cdot \psi \\ &= \psi_0^2 + \psi_g^2 + 2\psi_0\psi_g \cos(\mathbf{g} \cdot \mathbf{r}). \end{aligned} \qquad (4.23)$$

It is seen from Equation (4.22) and very clearly from Equation (4.23) that both field amplitude and intensity of the combination of incident and diffracted waves is periodic with respect to $\mathbf{g} \cdot \mathbf{r} = 2\pi$. Substitution for g using Equation (4.18) reveals that the wave is periodic with respect to the atomic spacing d. If we add to $\mathbf{r}$ a component perpendicular to the lattice planes of magnitude d, then the amplitude is left unchanged as $\mathbf{g} \cdot \mathbf{r}$ is simply incremented by 2π. In other words the combination of incident plane wave and diffracted plane wave is a Bloch wave. This is the type of wave we find inside the crystal. The electron does not propagate as a plane wave, but as a linear combination of plane waves whose wavevectors are linked by the Laue condition.

4.4 The nearly free electron model

Let us now consider a rather spcial case, that of Bragg reflection at normal incidence. Then the Bragg equation (4.17) becomes

$$n\lambda = 2d$$

or

$$k = n\pi/d. \tag{4.24}$$

At these values of k, the Bloch wave consists of two plane wave components propagating in opposite directions. This is a very special case because, instead of the electron wave being a travelling wave, the two components combine to form a standing wave.

For simplicity let us consider the specific case of a Bloch wave of wavevector $k = n\pi/a$ where a is the atomic spacing parallel to a cube edge direction x of a simple cubic crystal. The spacing a will then be the spacing on the nearest atomic planes perpendicular to the cube edge direction. In this case the Bloch wave has two plane wave components, one propagating in the $+x$ direction and the other propagating in the $-x$ direction. These combine to form standing waves and, depending on the relative phases between them, we have two possible solutions

$$\begin{aligned} \psi_+ &= \psi_0[\exp(\mathrm{i}kx) + \exp(-\mathrm{i}kx)] \\ &= 2\psi_0 \cos kx \end{aligned} \tag{4.25a}$$

and

$$\begin{aligned} \psi_- &= \psi_0[\exp(\mathrm{i}kx) - \exp(-\mathrm{i}kx)] \\ &= 2\mathrm{i}\psi_0 \sin kx. \end{aligned} \tag{4.25b}$$

By symmetry $|\psi_0| = |\psi_g|$.

The intensity of the wave is proportional to the electron density ρ and thus the negative charge density has two possible values. From Equations (4.25) we have directly,

$$\rho(+) = A\cos^2 kx \tag{4.26a}$$

$$\rho(-) = A\sin^2 kx \tag{4.26b}$$

where A is a proportionality constant.

Figure 4.7 illustrates how the two solutions for the electron density vary with distance x. We see that one solution concentrates electron density at the atomic planes, the other concentrates it mid-way between the atomic planes. In other words the solution $\rho(+)$ concentrates negative charge at the (positive) ion cores, while $\rho(-)$ concentrates negative charge away from the ion cores. Due to the Coulomb interaction, the electrostatic energy associated with the $\rho(+)$ solution is less than that associated with the $\rho(-)$ solution.

At these specific values of k, given by integral multiples of π/a we then have two possible energies for the electron wave. For each value of k there are two possible energies. The $E-k$ curve is no longer single valued. Let us recall the $E-k$ relation for the free electron, namely that the free electron energy E as a function of wavevector, is

$$E = \hbar^2k^2/2m \tag{4.27}$$

(Note $p = \hbar k$ and $E = p^2/2m$ where p is momentum.) Then the $E-k$ curve is a parabola (Fig. 4.8(a)) and is single valued. What happens

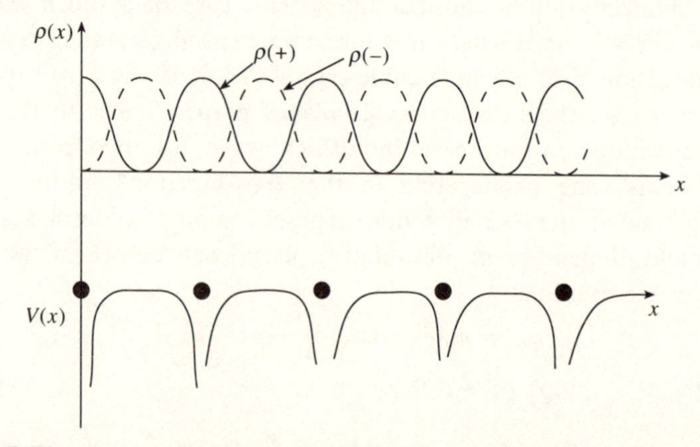

Fig. 4.7 The electron density at the Brillouin zone boundary. Two solutions exist $\rho(+)$ and $\rho(-)$ distributing electron density either closer to or further from the positive ion cores.

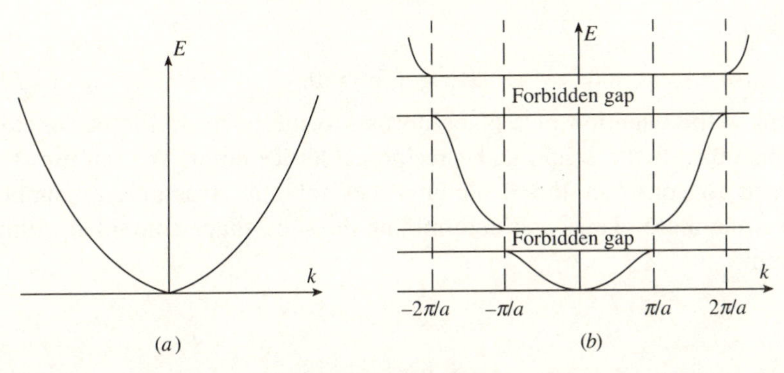

Fig. 4.8 (*a*) Energy–wavevector (E–k) diagram under the free electron approximation. (*b*) E–k diagram in the nearly free electron approximation showing the formation of forbidden energy gaps.

when we introduce the effect of a periodic perturbation on the potential? Then we have seen that at $k = n\pi/a$ the E–k curve is two valued. Thus, if the perturbation to the free electron potential is sufficiently small, i.e. the electrons are *nearly free*, the E–k curve takes the form shown in Fig. 4.8(*b*). Far from the values of $k = n\pi/a$, the E–k curve is almost parabolic. At $k = n\pi/a$ a discontinuity occurs. The size of the discontinuity represents the potential energy difference between the two solutions ψ_+ and ψ_-. As the total energy must be conserved, the kinetic energy must differ by the difference in potential energy.

This simple treatment has led us to a concept of crucial importance, namely that at a specific values of k, known as the Brillouin zone boundary (for reasons which will become clear with our subsequent study of solid state physics) there is a discontinuity in the E–k curve. Due to this discontinuity there exist values of electron energy E which are forbidden. There are energy gaps, values of energy which the electron cannot take. Conversely, we can say that because of these forbidden gaps, there are only *bands* of energy which the electron is allowed to take. We have imposed a restriction which will enable us to explain the difference in properties between metals, insulators and semiconductors.

The gaps, as we have noticed in this one-dimensional example, occur at values of the wavevector where Bragg reflection occurs. When we extend the discussion to three dimensions, we still find that the Brillouin zone boundaries exist where Bragg reflection occurs. If we square the Laue equation (4.16), we obtain

$$|\mathbf{k}_g|^2 = |\mathbf{k}_0|^2 + 2\mathbf{k}_0 \cdot \mathbf{g} + |\mathbf{g}|^2 \tag{4.28}$$

which is

$$2\mathbf{k}_0 \cdot \mathbf{g} + g^2 = 0. \qquad (4.29)$$

This is the equation of a plane normal to the reciprocal lattice vector which bisects the origin and a reciprocal lattice point. We can use this result to construct the Brillouin zones for any structure, by simply drawing these planes and determining the solid shape contained within them.

4.4.1 *Effective mass*

One of the more bizarre concepts which we now meet, predicted by the nearly free electron model, is that of 'effective mass'. We will find that the effective mass of an electron in a crystal is not the same as the mass of a free electron. The periodic crystal potential alters quite dramatically the way in which the electrons behave.

As we have seen, the wavefunction of a travelling plane wave propagating in the $+x$ direction can be written

$$\psi = \psi_0 \exp[\mathrm{i}(\omega t - kx)]. \qquad (4.30)$$

The observed group velocity v is given by

$$v = \partial\omega/\partial k \qquad (4.31)$$

and as

$$E = \hbar\omega \qquad (4.32)$$

we have

$$v = \frac{1}{\hbar}\left(\frac{\partial E}{\partial k}\right). \qquad (4.33)$$

Let us consider an electron of charge e under the influence of an electric field $\mathscr{E}$. In time $\mathrm{d}t$ it gains energy $\mathrm{d}E$ given by

$$\mathrm{d}E = e\mathscr{E}v\mathrm{d}t. \qquad (4.34)$$

Therefore

$$\frac{\partial E}{\partial t} = \frac{e\mathscr{E}}{\hbar}\left(\frac{\partial E}{\partial k}\right) \qquad (4.35)$$

using Equation (4.33). From Equation (4.35) we have

$$\frac{\partial k}{\partial t} = \frac{e\mathscr{E}}{\hbar} \qquad (4.36)$$

The acceleration $\partial^2 x/\partial t^2$ is given by

$$\frac{\partial^2 x}{\partial t^2} = \frac{\partial v}{\partial t} = \frac{\partial}{\partial t}\left(\frac{1}{\hbar}\frac{\partial E}{\partial k}\right)$$
$$= \frac{1}{\hbar}\frac{\partial}{\partial k}\left(\frac{\partial E}{\partial k}\right)\left(\frac{\partial k}{\partial t}\right)$$
$$= \frac{1}{\hbar}\left(\frac{\partial^2 E}{\partial k^2}\right)\left(\frac{\partial k}{\partial t}\right). \quad (4.37)$$

The force on the electron is $e\mathcal{E}$. If we define the effective mass as m^* then from Newton's Second Law we have

$$F = m^*(\partial^2 x/\partial t^2)$$
$$= e\mathcal{E}. \quad (4.38)$$

Combining Equations (4.36), (4.37) and (4.38) yields

$$m^* = \frac{\hbar^2}{(\partial^2 E/\partial k^2)}. \quad (4.39)$$

This is a specific case of a much more general tensor result.

Equation (4.39) yields no surprises for the case of a free electron. Using it in Equation (4.27) gives $m^* = m$. That is satisfactory. The effective mass is the free electron mass. However, in a periodic potential, the nearly free electron model does yield a startling result. Examination of Fig. 4.8(*b*) shows that it has been drawn so that just above and below the energy gaps, the gradient of the $E-k$ curve is zero. This is no accident. A few minutes sketching will show that in order to retain the parabolic shape far from the Brillouin zone boundaries and two values of energy at the zone boundaries, this is necessary. There is however another reason for this zero gradient. Equation (4.31) gives the group velocity v as proportional to $\partial E/\partial k$. In other words at the Brillouin zone boundaries, as $\partial E/\partial k = 0$, the group velocity is zero. This is just what we expect, because we have already deduced that the Bloch wave there is a standing wave, and standing waves have zero group velocity.

However, this leads to a very important observation. Just below the energy gap, just inside the Brillouin zone boundary where k is just below π/a for example, in Fig. 4.8(*b*), $\partial^2 E/\partial k^2 < 0$. In other words the effective mass m^* is *negative*.

4.4.2 Holes

The above result can come as something of a shock to some readers who, until this moment, thought that they understood what is meant by mass. Nevertheless when pressed, they usually admit that mass is defined as a proportionality constant relating force to acceleration. What is meant by negative mass? Essentially that the object accelerates in the opposite direction to the force. Snooker with negative mass balls could be a great spectator sport! In the present case we are concerned with electromagnetic forces. The direction of the force on a charged particle in an electric field depends on the sign of the charge of the particle. Therefore the acceleration can be either parallel or antiparallel to the direction of the force, depending on the sign of the charge. Fig. 4.9 shows the direction of acceleration for negatively and positively charged particles with positive and negative effective mass. Because

$$\mathrm{d}^2x/\mathrm{d}t^2 = e\mathcal{E}/m^* \tag{4.40}$$

it is quite clear that a positively charged particle with negative effective mass behaves just like a normal negatively charged electron with positive mass. On the other hand an electron with negative mass behaves just like a positively charged particle with positive effective mass. That is, an electron with $m^* < 0$ appears to carry a positive charge. We now begin to see the origin of the positive Hall coefficients in aluminium and indium. Because of the interaction with the crystal lattice potential, the conduction electrons appear to carry positive charge.

Field direction ←

	Acceleration direction		Acceleration direction
$e<0$ $m^*>0$	→	$e<0$ $m^*<0$	←
$e>0$ $m^*>0$	←	$e>0$ $m^*<0$	→

Fig. 4.9 Direction of acceleration for constant force with positive and negative charges and masses.

When the conduction electrons have negative effective mass, i.e. they appear to have positive charge we refer to them as *holes*. Consideration of the behaviour of electrons in a nearly filled band enables us to obtain further insight into the origin of this effect. Let us consider the limit of a very nearly filled band. In fact the ultimate limit is where electrons fill all available states in the band except one (Fig. 4.10(*a*)). When an electric field is applied, electrons close to the Fermi level (just below the band edge) can gain energy and accelerate. However as there is only one vacant state, excitation of one electron fills the band except for the state it has left. In other words a *hole* is left lower down in the band (Fig. 4.10(*b*)). Excitation of electrons can only take place by excitation into that hole, leaving a hole even lower down in the band. The effect of the electric field is to move the hole down the energy band. To all intents and purposes we can consider the behaviour of the single hole rather than all of the electrons. As the hole represents a gap in an otherwise uniform sea of negative charge, it looks like a single positive charge. Note once again that increasing hole energy corresponds to movement down the band, i.e. in the opposite sense to an electron.

A number of homely analogies exist which the student may find helpful to grasp this point. In a spirit level, used for building work, a bubble of liquid is used to determine whether a plane is horizontal. One does not consciously think of the liquid moving, attention is focussed on the movement of the bubble. When observing small bubbles of carbon dioxide rising in a glass of champagne or soda water, one does not consciously think of how the liquid is being displaced and flowing in order to fill the space left as the gas bubble rises. One tends to think of the bubble, even though the molecular density in the bubble is only a small fraction of that in the liquid. In effect we are

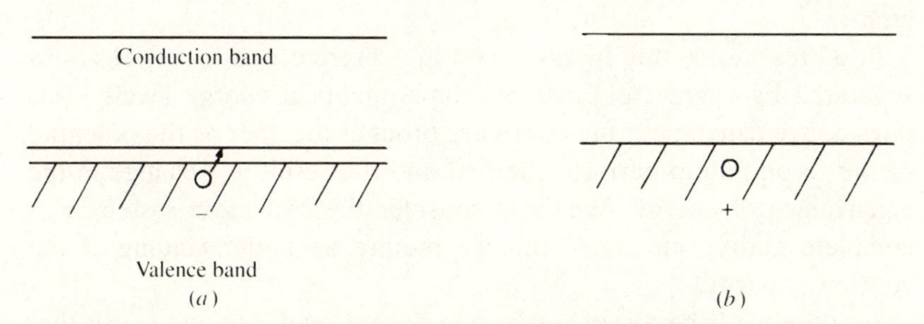

Fig. 4.10 Almost filled valence band showing how excitation into a vacant state can lead to formation of a hole lower down the band.

studying the behaviour of the hole in the liquid, not the behaviour of the liquid itself. It is clear that the analogy extends to the case of holes in an otherwise uniform electron sea.

One other point which can be made here is that because the mechanism of hole excitation occurs by excitation of electrons successively into the vacant hole state, the mobility of holes is rather low. This result will be used later in our discussion of semiconductors. An analogy which provides for some mirth in the lecture room is to demonstrate the ease with which a person can move along an empty row of seats, compared with the difficulty of moving a whole row of people so that an empty space in an almost full row of seats can be transferred down the row.

4.5 The tightly bound electron model

We have considered the effect of a small periodic perturbation on the free electron model, and to a good approximation this nearly free electron model accounts rather well for the properties of the conduction electrons. However, in our present model there is only one conduction electron per atom, and that applies to metals only. Clearly we must extend our model to account for the properties of insulators, even if no more than to explain simply why some materials are metals and others insulators. In order to do this we must examine the electronic configuration of the free atoms and study how this is changed as the atoms are brought together to form a solid. This approach assumes that the electrons remain sufficiently tightly bound to the atoms of the solid to be identified with quantum states in the free atom. To a good approximation this tightly bound model is valid for the core electrons while the nearly free model describes the conduction electrons (which correspond to the valence electrons in the atom).

In a free atom, the energy levels are discrete and identical atoms separated by a great distance will have identical energy levels. This does not remain true if the atoms are brought together as the potential of the second atom perturbs the first and the result is a change in the eigenvalues of energy. We must consider the two atom system as a complete entity, and to do this we require an understanding of the concept of parity.

Suppose we have a symmetric, one-dimensional, potential such that $V(x) = V(-x)$. In this case Schrödinger's equation will be unchanged by substitution of x for $-x$. One might then expect that the solutions would be identical, that is $\psi(x) = \psi(-x)$. However, an important tenet of quantum mechanics is that we cannot measure the amplitude of the wavefunction, only the probability density $|\psi(x)|^2$. Therefore we can only say that the symmetry requires

$$|\psi(x)|^2 = |\psi(-x)|^2. \tag{4.41}$$

This implies

$$\psi(x) = \psi(-x) \tag{4.42a}$$

or

$$\psi(x) = -\psi(-x). \tag{4.42b}$$

The first type of solution is said to have even parity, the second odd parity. It is important to stress that any solution must have a definite parity, a point which will have implications later.

Consider the simple case of a single, infinitely deep, one-dimensional, potential well. The ground state wavefunction is sketched in Fig. 4.11(a). It has even parity. (The reader can verify that the first excited state has odd parity.) Now consider the case of two infinitely deep potential wells A and B, as shown in Fig. 4.11(b). There are only two possibilities for the total ground state wavefunctions which have a definite parity. The total wavefunction can be either

$$\psi_+ = \frac{1}{\sqrt{2}}(\psi_A + \psi_B) \tag{4.43a}$$

or

$$\psi_- = \frac{1}{\sqrt{2}}(\psi_A - \psi_B). \tag{4.43b}$$

Because there is no communication between the wells, these two states have the same energy and are said to be degenerate.

Two non-communicating wells correspond to the situation where atoms are separated by a large distance and the energy levels remain that of the isolated system. If we now allow communication between wells (Fig. 4.11(c)) we get to the situation corresponding to the atoms being brought together. Now the two possible wavefunctions are different and most importantly have different curvature. The solution ψ_- has more curvature in the central region than ψ_+ and hence the solution ψ_- has greater energy than ψ_+. As a result of the breaking of the degeneracy, the discrete single well energy levels have been split into doublets.

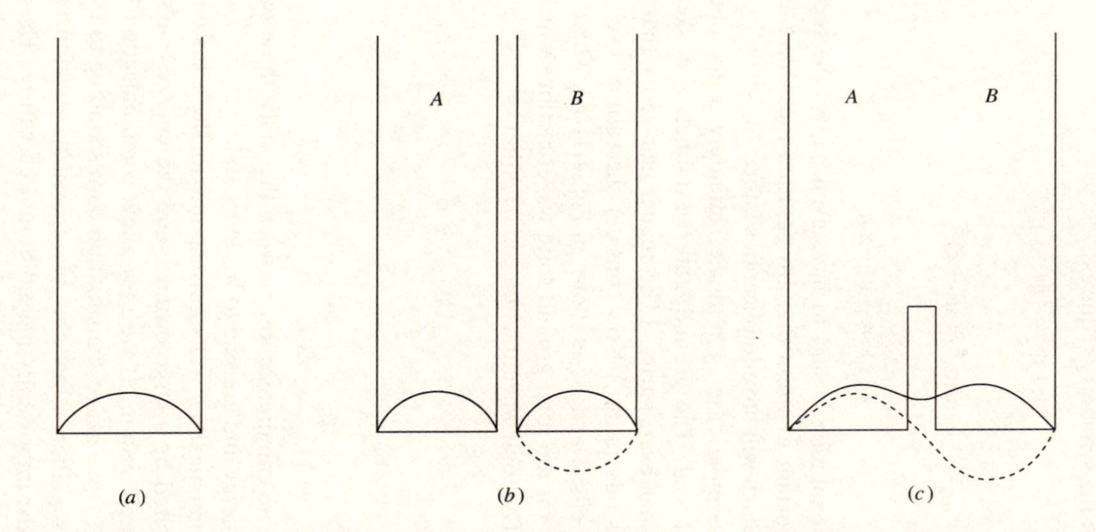

Fig. 4.11 Infinitely deep one-dimensional quantum wells. (*a*) Single well (*b*) two non-communicating wells (*c*) two wells with finite barrier between them. The solid line is the solution for ψ_+, the dashed line for ψ_-.

We can follow this process on by adding successive atoms. As more and more atoms are added, so the discrete atomic levels split up into more and more levels. Indeed, as illustrated in Fig. 4.12, for an ensemble of N atoms, each discrete atomic level splits into N sub-levels. The result is a series of bands, each containing N very closely spaced levels, separated by an energy gap. We have arrived once again, by a totally different route, at a concept of a quasi-continuum of allowed states in bands separated by forbidden gaps.

4.5.1 Formal description of the tightly bound model

We will now take these ideas further in order to obtain, for a one-dimensional linear chain, a relation between the electron energy E and the wavevector k under the tightly bound electron approximation. Suppose an electron is simply attached to an atom labelled n in the linear chain. In this case, we can solve the *time dependent* Schrödinger equation by separating the time and space variables, $\Psi_n(x, t) = \psi_n(x)\Phi_n(t)$, which gives, for the time dependent part

$$\mathrm{i}\hbar \mathrm{d}\Phi_n(t)/\mathrm{d}t = E_0\Phi_n(t) \tag{4.44}$$

which has a solution

$$\Phi_n(t) = \exp(-\mathrm{i}E_0 t/\hbar). \tag{4.45}$$

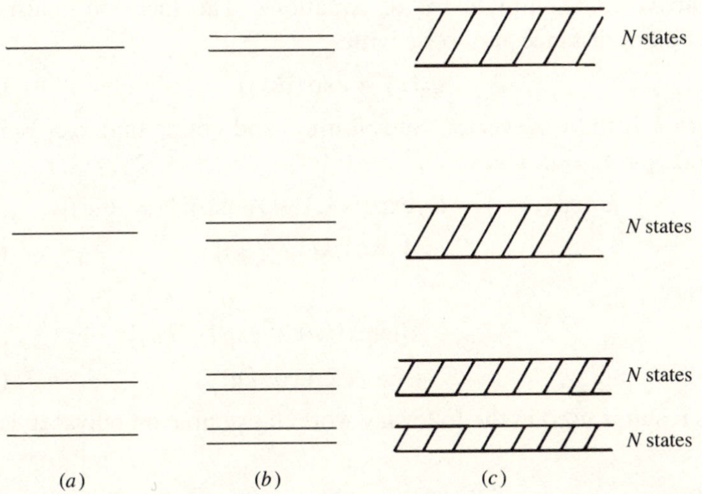

N states

N states

N states

N states

(*a*) (*b*) (*c*)

Fig. 4.12 Splitting of single atom levels (*a*) into doublets as two atoms are brought together (*b*); with N interacting atoms (*c*) each original level becomes a quasi-continuous band containing N levels.

We can go on to solve the spatial part of the wavefunction which is independent of time. This could have been done for any atom and there will be equivalent solutions for atoms $n + 1$, $n - 1$, etc.

Now let us consider what happens if we let an electron 'leak' from atom number n to nearest-neighbour atoms $n + 1$ and $n - 1$. There is no well defined 'eigenfunction' corresponding to Φ and no definite 'eigenvalue' of energy E_0. However, we can find a solution of the Schrödinger equation as a linear combination of all such eigenfunctions. In this case that is saying that the wavefunction is a linear combination of the wavefunctions of electrons attached to specific atoms. The Hamiltonian equation now becomes

$$\mathrm{i}\hbar \mathrm{d}\Phi_n(x, t)/\mathrm{d}t = E_0\Phi_n(x, t) - A\Phi_{n+1}(x, t) - A\Phi_{n-1}(x, t), \quad (4.46)$$

if we assume that the electron can only leak to nearest neighbours and the probability amplitude for this to happen is A. Note that A and E_0 are time independent.

The argument could have just as easily been made for an electron at atom $n + 1$ or $n - 1$ and thus, equivalently, we have

$$\mathrm{i}\hbar \mathrm{d}\Phi_{n-1}(x, t)/\mathrm{d}t = E_0\Phi_{n-1}(x, t) - A\Phi_n(x, t) - A\Phi_{n-2}(x, t) \quad (4.47)$$

$$\mathrm{i}\hbar \mathrm{d}\Phi_{n+1}(x, t)/\mathrm{d}t = E_0\Phi_{n+1}(x, t) - A\Phi_{n+2}(x, t) - A\Phi_n(x, t). \quad (4.48)$$

If we look for a solution of the form

$$\Phi_n = q_n \exp(-\mathrm{i}Et/\hbar) \quad (4.49)$$

we arrive at an infinite set of equations. The factor q_n must be a function of distance and so we write

$$q_n(x) = \exp(\mathrm{i}kx_n) \quad (4.50)$$

where k is the wavevector. Substituting, and noting that $x_{n+1} = x_n + a$ and $x_{n-1} = x_n - a$, gives

$$E \exp(\mathrm{i}kx_n) = E_0 \exp(\mathrm{i}kx_n) - A \exp[\mathrm{i}k(x_n + a)] - A \exp[\mathrm{i}k(x_n - a)]. \quad (4.51)$$

Thus

$$E = E_0 - A[\exp(\mathrm{i}ka) + \exp(-\mathrm{i}ka)] \quad (4.52)$$

$$E = E_0 - 2A \cos ka. \quad (4.53)$$

This result is used in the following worked example on polyacetylene.

Worked example

Polyacetylene is one of a new class of conducting polymers and has been the subject of extensive research in the Chemistry Department both at Durham University and various other establishments. Its structure is

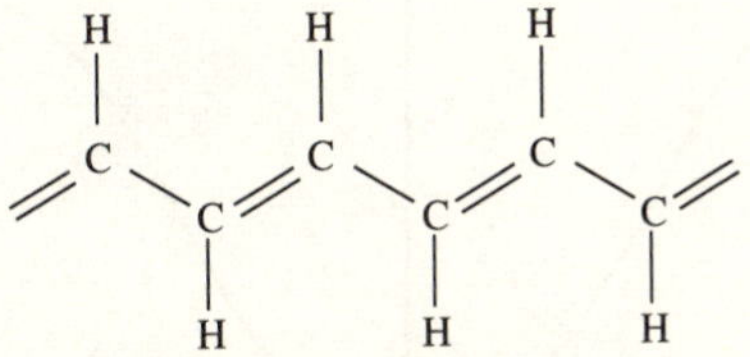

It is possible to free an electron from the double bond without altering the polymer backbone and it should behave like a one-dimensional metal. The ideal band structure is of a conduction band with energy E given by

$$E = E_0 - 2A \cos ka$$

where k is the electron wavevector and a is the carbon–carbon distance.

(*a*) Sketch the E–k curve in the first Brillouin zone.
(*b*) Determine the electron effective mass at the zone boundary.
(*c*) Determine the effective mass at E_0.

Now the Fermi energy coincides with E_0 and to avoid the above problem, polyacetylene dimerizes, i.e. the carbon–carbon distance is no longer equal in the double and single bonds (this is known as a Peierls distortion) and a band gap appears at the new Brillouin zone boundary.

(*d*) Sketch the new band structure in the first two Brillouin zones and show clearly the size of the zones.

Solution

(*a*) The E–k curve is shown in Fig. 4.13(*a*). We see that the zone boundaries occur at wavevector $k = \pm\pi/a$ and that the width of the energy band is $4A$.
(*b*) The effective mass is given in Equation (4.39) as

$$m^* = \hbar^2/(\partial^2 E/\partial k^2).$$

Therefore, as

$$\partial E/\partial k = 2aA \sin ka,$$

we have

$$\partial^2 E/\partial k^2 = 2a^2 A \cos ka.$$

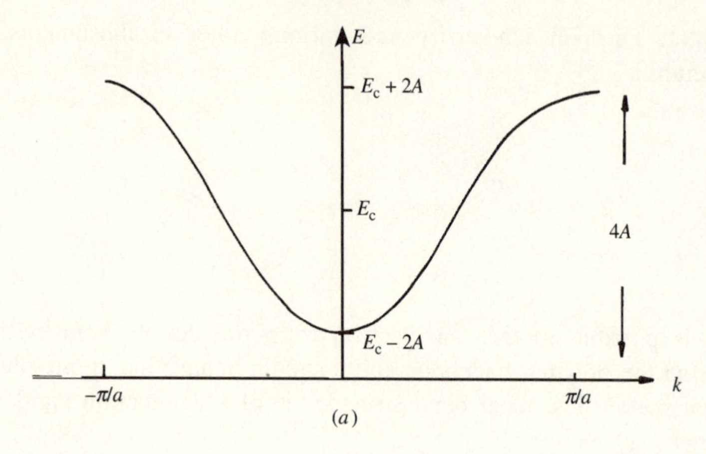

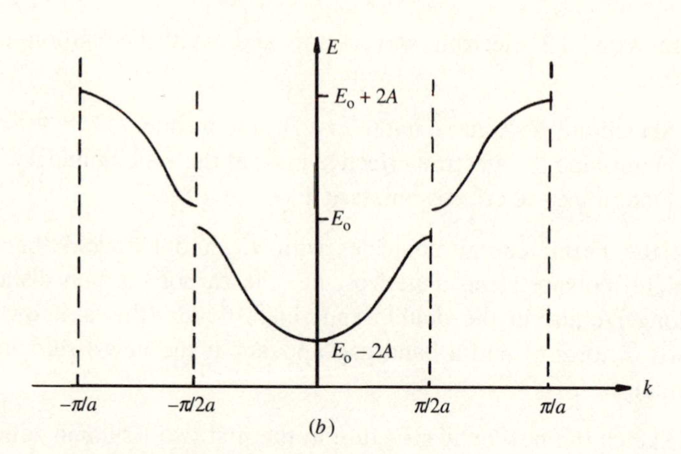

Fig. 4.13 (*a*) Band structure of polyacetylene assuming all C–C bonds identical. (*b*) Band structure of dimerized state.

Thus

$$m^* = \hbar^2/(2a^2 A \cos ka).$$

At the zone boundary $k = \pm\pi/a$ and

$$m^* = -\hbar^2/(2a^2 A).$$

(c) For $E = E_0$, we find $\cos ka = 0$. Thus at E_0, $m^* \to \infty$.

(d) The new band structure is sketched in Fig. 4.13(*b*). As the C—C bond is now not equal to the C=C bond in length, the unit cell is doubled. The repeat distance is now $2a$, not a. Therefore the first Brillouin zone boundary occurs at $k = \pm\pi/2a$.

The Fermi energy now lies in the middle of an energy gap and therefore (undoped) polyacetylene behaves like a semiconductor, rather than a metal.

4.6 General approach to energy bands

We saw qualitatively that energy gaps occur at the Brillouin zone boundaries, e.g. at wavevector $k = \pm\pi/a$ for the case of a one-dimensional lattice of period a. For this case, let the crystal potential be $V(x)$ where, of course, $V(x + a) = V(x)$. As this is a periodic function we may expand this as Fourier series. This is done over the reciprocal lattice vectors **g**. (The periodicity in k space is **g** where in real space it is a. From Equation (4.18) we see that here $g = 2\pi/a$.) Thus,

$$V(x) = \sum_g V_g \exp(igx). \tag{4.54}$$

Although we must sum over all g values, the V_g values decrease very rapidly with magnitude of g. In the three-dimensional case this generalizes very readily to

$$V(\mathbf{r}) = \sum_g V_g \exp(i\mathbf{g}\cdot\mathbf{r}) = (\hbar^2/2me)\sum_g U_g \exp(i\mathbf{g}\cdot\mathbf{r}). \tag{4.55}$$

Now we need to solve the Schrödinger equation

$$(\hbar^2/2m)\nabla^2\psi(\mathbf{r}) + [E - eV(\mathbf{r})]\psi(\mathbf{r}) = 0 \tag{4.56}$$

where E is the electron kinetic energy. We must look for solutions which have a periodicity equal to that of the crystal lattice, in order to satisfy Bloch's theorem. A possible solution is a Bloch wave of the form

$$\psi(\mathbf{r}) = \sum_g C_g \exp[i(\mathbf{k} + \mathbf{g})\cdot\mathbf{r}]. \tag{4.57}$$

(Equation (4.57) is an example of a function satisfying the general requirement of the Bloch theorem, and is of the form

$$\psi_k(\mathbf{r}) = u_k(\mathbf{r})\exp(i\mathbf{k}\cdot\mathbf{r}) \tag{4.58}$$

where $u_k(\mathbf{r})$ has the period of the crystal lattice. This can be described formally as

$$u_k(\mathbf{r}) = u_k(\mathbf{r} + \mathbf{T}) \tag{4.59}$$

where **T** is a lattice translation vector, which means that points at **r** and **r** + **T** are in every way equivalent.)

We now substitute Equations (4.55) and (4.57) in (4.56). In doing so we note that

$$V(\mathbf{r})\psi(\mathbf{r}) = (\hbar^2/2me)\sum_g U_g \sum_h C_h \exp[\mathrm{i}(\mathbf{k}+\mathbf{g}+\mathbf{h})\cdot\mathbf{r}]$$

$$= (\hbar^2/2me)\sum_g \exp[\mathrm{i}(\mathbf{k}+\mathbf{g})\cdot\mathbf{r}]\left(U_0 C_g + {\sum_h}' U_h C_{g-h}\right) \quad (4.60)$$

which is obtained by rearranging the terms in the double summation over reciprocal lattice vectors labelled **g** and **h**. The prime on the summation indicates that the **h** = 0 term is excluded.

We then obtain an infinite set of equations of the form

$$\exp[\mathrm{i}(\mathbf{k}+\mathbf{g})\cdot\mathbf{r}\{[(2m/\hbar^2)E - (\mathbf{k}+\mathbf{g})^2]C_g - {\sum_h}' U_h C_{g-h}\}] = 0 \quad (4.61)$$

when we set $U_0 = 0$. The exponential terms relate to different reciprocal lattice vectors and are independent. Their coefficients therefore can be independently equated to zero. Thus we have for each **g** value

$$[(\mathbf{k}+\mathbf{g})^2 - (2m/\hbar^2)E]C_g + {\sum_h}' U_h C_{g-h} = 0. \quad (4.62)$$

This equation is often called the 'central equation'. (It is central to electron diffraction theory too, . . . the only difference from band theory being the relative magnitudes of V and E.) Note that if the lattice potential vanishes, all C_g vanish except C_0 and thus we are left with $\psi(\mathbf{r}) = \exp(\mathrm{i}\mathbf{k}\cdot\mathbf{r})$ as for a free electron.

4.6.1 Solution of the central equation

For a non-trivial solution of Equation (4.62), the determinant of the coefficients must vanish. While this is, in principle, a determinant of infinite extent, we see that the U_g coefficients become very small for large g and hence certain approximations may be made which can then be handled.

4.6.2 The empty lattice approximation

If we assume that the crystal potential becomes vanishingly small we can approximate all $U_g = 0$. Then the solution of Equation (4.62) is

$$E = (\hbar^2/2m)(\mathbf{k}+\mathbf{g})^2. \quad (4.63)$$

Note that this is very similar to the free electron solution, but it provides a valuable means of illustrating the conceptual tools used in band theory. We have here a periodic potential which is allowed to become vanishingly small. However, the essential periodic nature of the structure is retained and this has far-reaching consequences for the behaviour of electrons in solids.

Reduced zone scheme

Because any vector $\mathbf{k}'$ which is larger than a reciprocal lattice vector $\mathbf{g}$ can be represented by a shorter vector $\mathbf{k}$ INSIDE the first Brillouin zone by the transformation

$$\mathbf{k}' = \mathbf{k} + \mathbf{g}, \tag{4.64}$$

only the first Brillouin zone is used in an energy band diagram. All the wavevectors and associated energies are 'folded back' into the first zone (Fig. 4.14).

This equivalence of waves with wavevectors $\mathbf{k}$ and $\mathbf{k} + \mathbf{g}$ can be understood most readily by consideration of a displacement wave of a line of beads coupled together by elastic thread. As seen in Fig. 4.15, one can create an infinite number of waves which will describe the displacement of the beads. However, the only physical description

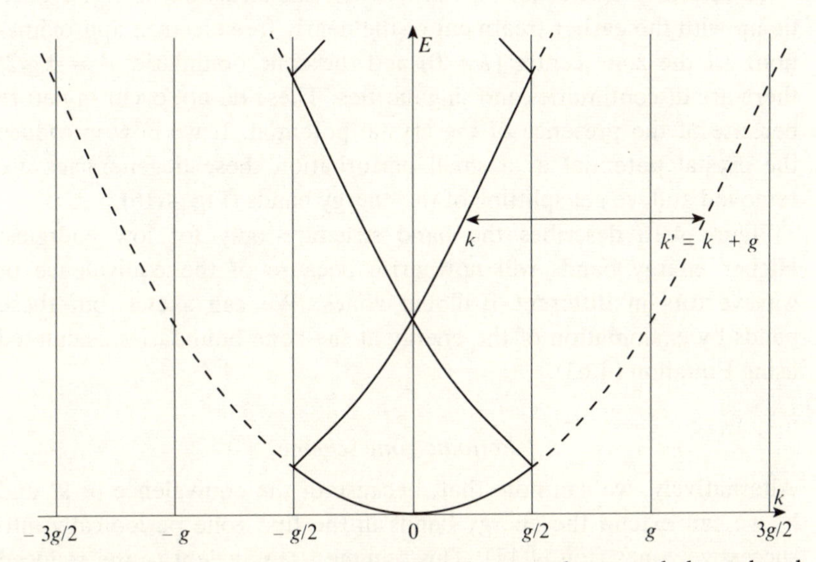

Fig. 4.14 The relation between the extended zone scheme and the reduced zone scheme where parts of the E–k curve outside the first Brillouin zone are folded back into it.

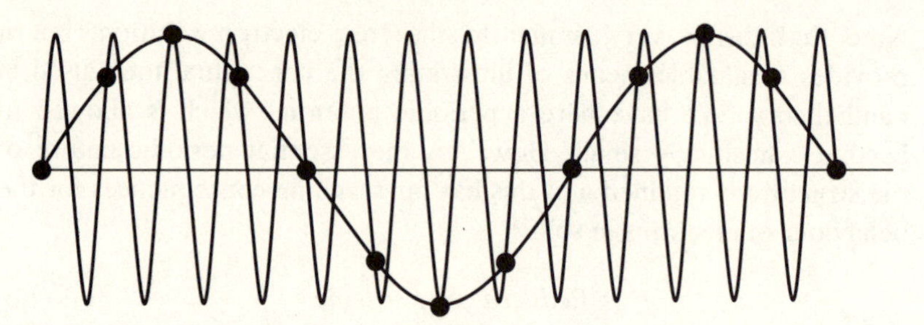

Fig. 4.15 Displacement of beads on an elastic string. The high frequency wave (9ω) contains no more physical information than the fundamental wave of frequency ω.

which is meaningful is the lowest wavelength wave. All others have higher wavevector and have many undulations which do not correspond to atomic positions. It is easy to see that for a line of beads of spacing a, the wavevector at which this duplication sets in is when $k = \pi/a$. Here, alternate beads are oscillating in antiphase, and this is the Brillouin zone boundary for the vibration of the beads. This analogy is *directly* relevant to the study of lattice vibrations in crystals where similar concepts apply to those being developed for electrons.

To return to electrons, we can now use the structure of Fig. 4.14 to tie up with the earlier treatment of the nearly free electron approximation. At the zone centre ($k = 0$) and the zone boundaries $k = \pm g/2$, there are discontinuities and singularities. These do not occur in nature because of the presence of the crystal potential. If we now introduce the crystal potential as a small perturbation these degeneracies are removed and we get splitting of the energy bands (Fig. 4.16).

Figure 4.16 describes the band structure only for low energies. Higher energy bands will now arise because of the equivalence of wavevectors in different Brillouin zones. We can sketch out these bands by examination of the energy at the zone boundaries calculated using Equation (4.63).

Periodic zone scheme

Alternatively, we can note that, because of the equivalence of $\mathbf{k}'$ and $\mathbf{k}$, we can extend the energy bands in the first zone periodically into successive zones (Fig. 4.17). This scheme is equivalent to the reduced zone scheme and used where it makes interpretation of the physics easier!

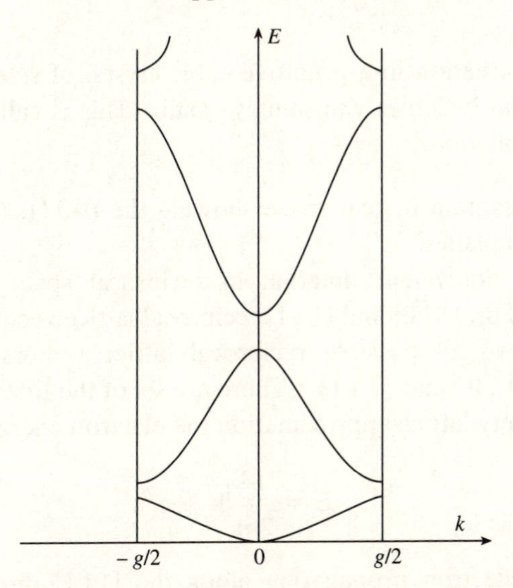

Fig. 4.16 Reduced zone scheme equivalent to Fig. 4.14, with degeneracies removed by the influence of the crystal potential. Band gaps naturally arise at the zone boundary and zone centre.

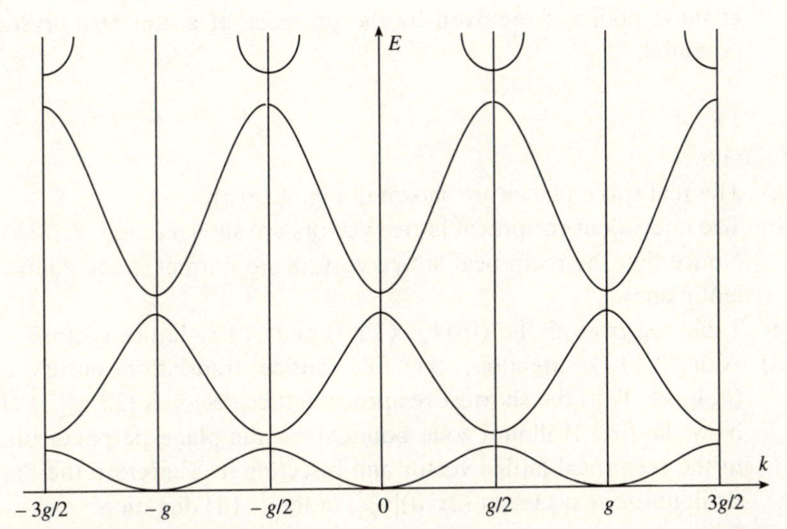

Fig. 4.17 Equivalent band diagram to that shown in Fig. 4.16, plotted in the periodic zone scheme.

Worked example

Consider the situation in a primitive cubic crystal of side a where the crystal potential becomes vanishingly small. (This is called the empty lattice approximation.)

(a) Draw a diagram in real space showing the (0 0 1), (1 1 0), $(1\,\bar{1}\,0)$, and (1 1 1) planes.

(b) Draw an equivalent diagram in reciprocal space showing the [0 0 1], [1 1 0], $[1\,\bar{1}\,0]$ and [1 1 1] reciprocal lattice vectors.

(c) Write down all possible reciprocal lattice vectors of the type $\langle 0\,0\,1 \rangle$, $\langle 1\,1\,0 \rangle$ and $\langle 1\,1\,1 \rangle$. (There are six of the first etc.)

(d) In the empty lattice approximation the electron energy is

$$E = \frac{\hbar^2}{2m}|\mathbf{k} + \mathbf{g}|^2.$$

Consider an electron propagating along the [1 1 1] direction in real space. It has then wavevector components $k_x = k_y = k_z = k/\sqrt{3}$. Prove that, in this direction, the first Brillouin zone boundary occurs at $(2\pi/a)[\frac{1}{2}\frac{1}{2}\frac{1}{2}]$.

(e) Sketch the eight lowest energy bands on a reduced zone scheme E–k plot. Ensure that the values are clearly identified for $k = 0$ and $\mathbf{k} = \pm[(\pi/a)\,(\pi/a)\,(\pi/a)]$.

(f) Sketch the shape of the lowest three bands when the degeneracy at these points is removed by the presence of a non-zero crystal potential.

Solution

(a) The real space planes are shown in Fig. 4.18(a).

(b) The equivalent reciprocal lattice vectors are shown in Fig. 4.18(b). Notice that the reciprocal lattice vectors are normal to the equivalent planes.

(c) Table 4.2 gives all the $\langle 0\,0\,1 \rangle$, $\langle 1\,1\,0 \rangle$ and $\langle 1\,1\,1 \rangle$ lattice vectors.

(d) Along [1 1 1] direction, the first lattice translation occurs at (a, a, a). Thus the shortest reciprocal lattice vector is $(2\pi/a)[1\,1\,1]$. Now the first Brillouin zone boundary is the plane perpendicular to the reciprocal lattice vector and bisecting it. Therefore the first Brillouin zone occurs at $(2\pi/a)[\frac{1}{2}\frac{1}{2}\frac{1}{2}]$ in the [1 1 1] direction.

(e) Table 4.3 shows the energies of an electron, propagating along [1 1 1] in real space, for the zone centre, general value of $\mathbf{k}$ and at

Table 4.2 *Reciprocal lattice vectors of the type ⟨001⟩, ⟨110⟩ and ⟨111⟩*

$[001], [00\bar{1}], [100], [\bar{1}00], [010], [0\bar{1}0]$
$[110], [1\bar{1}0], [\bar{1}\bar{1}0], [\bar{1}10], [101], [\bar{1}0\bar{1}], [10\bar{1}], [\bar{1}01]$
$[011], [0\bar{1}\bar{1}], [0\bar{1}1], [01\bar{1}]$
$[111], [\bar{1}\bar{1}\bar{1}], [11\bar{1}], [\bar{1}\bar{1}1], [1\bar{1}\bar{1}], [\bar{1}11], [\bar{1}1\bar{1}], [1\bar{1}1]$

the zone boundary. These data are used to sketch the relevant energy bands in Fig. 4.18(*c*).

(f) Fig. 4.18(*d*) shows the effect of a non-zero crystal potential in removing the degeneracies.

4.6.3 Solution near a zone boundary

Let us assume that the crystal potential is such that only the first Fourier coefficient U_g is important. (This is not too bad even for an unscreened Coulomb potential the U_g values fall off as g^{-2} and in practice it is faster.) Then the central equations become

$$[(\mathbf{k}+\mathbf{g})^2 - (2m/\hbar^2)E]C_g + U_{-g}C_0 = 0$$
$$[(\mathbf{k})^2 - (2m/\hbar^2)E]C_0 + U_gC_g = 0. \tag{4.65}$$

As only two C's are non-zero we have for the wavefunction

$$\psi(\mathbf{r}) = C_0 \exp(\mathrm{i}\mathbf{k}\cdot\mathbf{r}) + C_g \exp[\mathrm{i}(\mathbf{k}+\mathbf{g})\cdot\mathbf{r}] \tag{4.66}$$

which is a two component Bloch wave. In a centrosymmetric crystal structure

$$U_g = U_{-g} \tag{4.67}$$

and thus for a non-trivial solution

$$\begin{vmatrix} [(\mathbf{k}+\mathbf{g})^2 - (2m/\hbar^2)E] & U_g \\ U_g & [(\mathbf{k})^2 - (2m/\hbar^2)E] \end{vmatrix} = 0. \tag{4.68}$$

There are two solutions for the energy

$$E = \frac{\hbar^2}{4m}(k^2 + (k+g)^2 \pm \{[k^2 - (k+g)^2]^2 + 4U_g^2\}^{1/2}). \tag{4.69}$$

If we rewrite in terms of $\delta K = k - g/2$ we find after some manipulation

$$E = E(\pm g/2) + (\hbar^2/2m)\delta K^2(1 \pm g^2/2U_g). \tag{4.70}$$

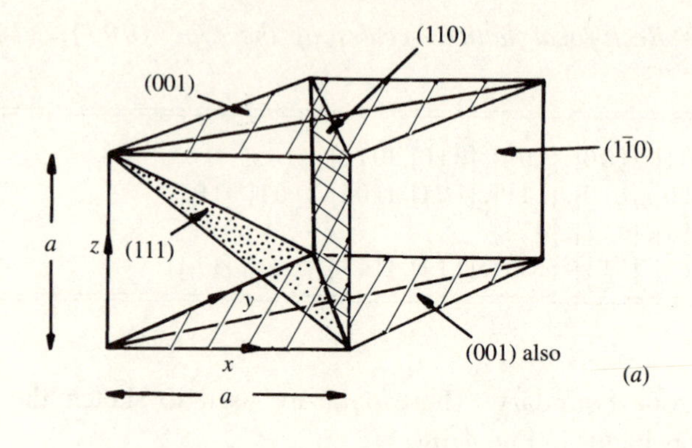
(110)
(001)
(1$\bar{1}$0)
a
z
(111)
y
(001) also
x
a
(a)

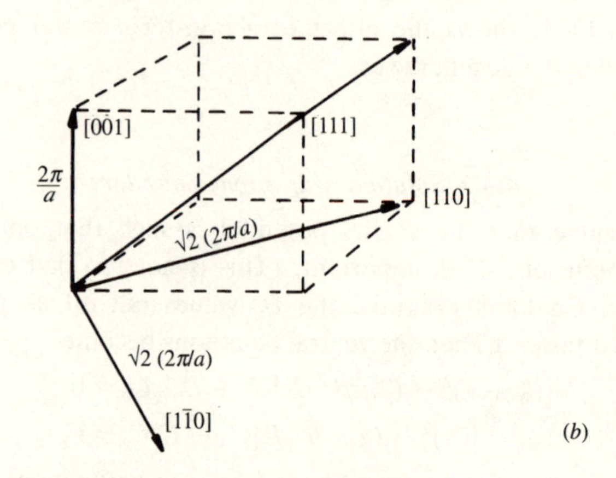
[001]
[111]
$\frac{2\pi}{a}$
[110]
√2 (2π/a)
√2 (2π/a)
[1$\bar{1}$0]
(b)

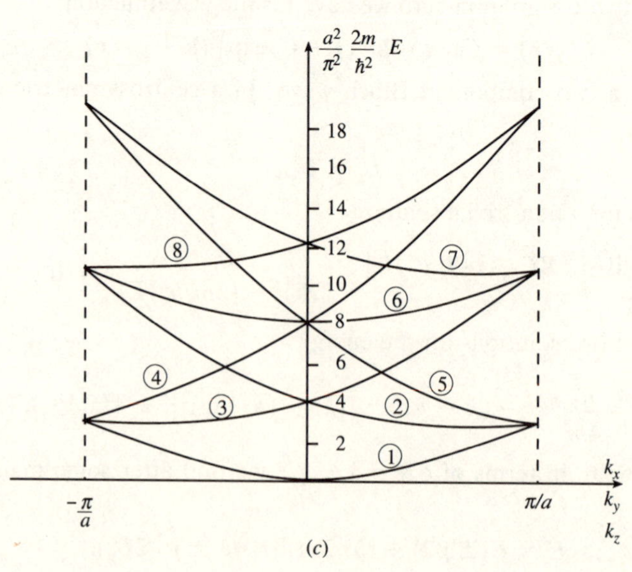
$\frac{a^2}{\pi^2}\frac{2m}{\hbar^2}E$
18
16
14
12
10
8
6
4
2
⑧
⑦
⑥
④
⑤
③
②
①
$-\frac{\pi}{a}$
π/a
k_x
k_y
k_z
(c)

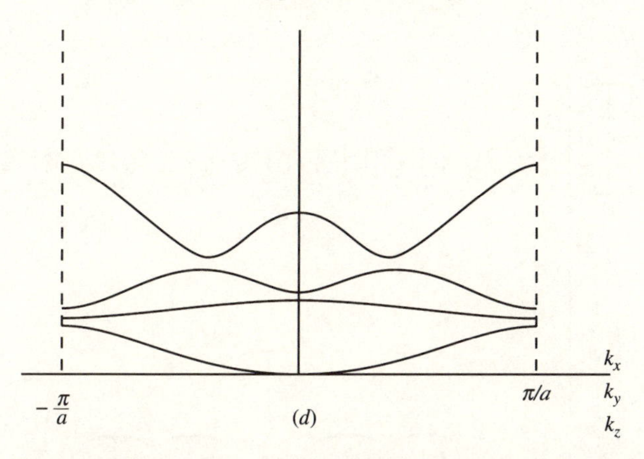

Fig. 4.18 (*a*) Real space planes (*b*) Equivalent reciprocal lattice vectors. (*c*) Lowest lying bands for electron propagation along the [111] direction of a cubic crystal calculated using the empty lattice approximation. (*d*) Equivalent diagram with degeneracy removed by non-zero crystal potential.

It is easy to see that the bands are parabolic close to the zone boundary. (You could get this result by expanding about a Taylor series in the periodic zone scheme by remembering that the $E-k$ curve is horizontal to the zone boundary as there the group velocity must be zero!) You need Equation (4.64), however, to see that the two solutions at the zone boundary ($k = g/2$ or $k = -g/2$) are

$$E(\pm g/2) = (\hbar^2/2m)(g^2/4 \pm U_g). \qquad (4.71)$$

This shows that the energy gap is $2eV_g$, i.e. a crystal potential of $2V_g \cos(gx)$ creates an energy gap of $2eV_g$.

4.6.4 The Kronig–Penney model

This one-dimensional model, sketched in Fig.4.19, assumes that the potential is a square periodic barrier of height V_0 and width b. The well width is a, giving a periodicity of $a + b$.

We need to solve Schrödinger's equation

$$(-\hbar^2/2m)\mathrm{d}^2\psi(x)/\mathrm{d}x^2 = [E - eV(x)]\psi(x) \qquad (4.72)$$

where $V(x) = 0$ for $0 < x < a$ and $V(x) = V_0$ for $-b < x < 0$.

Let us define

$$\alpha^2 = 2mE/\hbar^2 \quad \beta^2 = 2m(eV_0 - E)/\hbar^2. \qquad (4.73)$$

Table 4.3 *Electron energies in the empty lattice approximation for the lowest bands. Electron propagation along [1 1 1]*

Band	$g(a/2\pi)$	$\frac{2m}{\hbar^2}E(000)$	$\frac{2m}{\hbar^2}E(k_xk_yk_z)$	$\frac{2m}{\hbar^2}E\left(\frac{\pi}{a}\frac{\pi}{a}\frac{\pi}{a}\right)$
1	0	0	$3k_x^2$	$3\frac{\pi^2}{a^2}$
2	$[\bar{1}00]\,[00\bar{1}]\,[0\bar{1}0]$	$\left(\frac{2\pi}{a}\right)^2$	$2k_x^2+\left(k_x-\frac{2\pi}{a}\right)^2$	$3\frac{\pi^2}{a^2}$
3	$[100]\,[001]\,[010]$	$\left(\frac{2\pi}{a}\right)^2$	$2k_x^2+\left(k_x+\frac{2\pi}{a}\right)^2$	$11\frac{\pi^2}{a^2}$
4	$[1\bar{1}0]\,[10\bar{1}\,[01\bar{1}]$	$2\left(\frac{2\pi}{a}\right)^2$	$k_x^2+\left(k_x+\frac{2\pi}{a}\right)^2+\left(k_x-\frac{2\pi}{a}\right)^2$	$11\frac{\pi^2}{a^2}$
5	$[\bar{1}\bar{1}0]\,[\bar{1}0\bar{1}]\,[0\bar{1}\bar{1}]$	$2\left(\frac{2\pi}{a}\right)^2$	$k_x^2+2\left(k_x-\frac{2\pi}{a}\right)^2$	$3\frac{\pi^2}{a^2}$
6	$[110]\,[101]\,[011]$	$2\left(\frac{2\pi}{a}\right)^2$	$k_x^2+2\left(k_x+\frac{2\pi}{a}\right)^2$	$19\frac{\pi^2}{a^2}$
7	$[1\bar{1}\bar{1}]\,[\bar{1}\bar{1}1]\,[\bar{1}1\bar{1}]$	$3\left(\frac{2\pi}{a}\right)^2$	$2\left(k_x-\frac{2\pi}{a}\right)^2+\left(k_x+\frac{2\pi}{a}\right)^2$	$11\frac{\pi^2}{a^2}$
8	$[\bar{1}11]\,[1\bar{1}1]\,[11\bar{1}]$	$3\left(\frac{2\pi}{a}\right)^2$	$2\left(k_x+\frac{2\pi}{a}\right)^2+\left(k_x-\frac{2\pi}{a}\right)^2$	$19\frac{\pi^2}{a^2}$

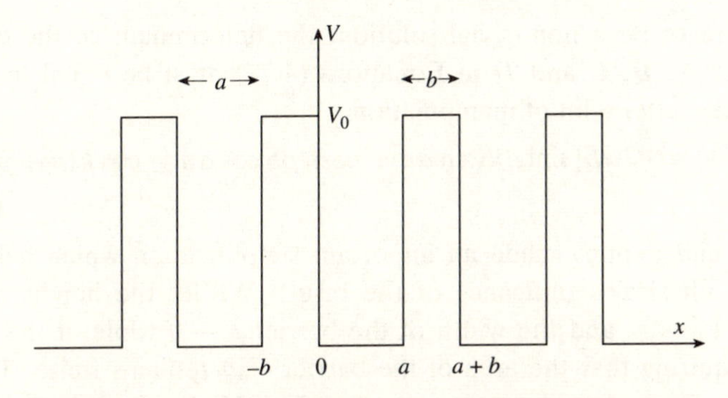

Fig. 4.19 The Kronig–Penney one-dimensional potential.

We look for solutions which are Bloch functions in the form $u_k(x)\exp(\mathrm{i}kx)$ and on substitution into Equation (4.72) we obtain

$$d^2u/dx^2 + 2ik(du/dx) + (\alpha^2 - k^2)u = 0 \quad \text{for } 0 < x < a \tag{4.74a}$$

$$d^2u/dx^2 + 2ik(du/dx) + (\beta^2 + k^2)u = 0 \quad \text{for } -b < x < 0 \tag{4.74b}$$

These equations have solutions of the form

$$u_1 = A\exp[\mathrm{i}(\alpha - k)x] + B\exp[-\mathrm{i}(\alpha + k)x] \quad \text{for } 0 < x < a \tag{4.75a}$$

$$u_2 = C\exp[(\beta - \mathrm{i}k)x] + D\exp[-(\beta + \mathrm{i}k)x] \quad \text{for } -b < x < 0 \tag{4.75b}$$

where A, B, C and D are constants. These must satisfy the conditions of continuity of the wavefunction and its derivative at all points, namely that

$$\begin{aligned} u_1(0) &= u_2(0) & u_1(a) &= u_2(-ba) \\ (du_1/dx)_{x=0} &= (du_2/dx)_{x=0} & (du_1/dx)_{x=a} &= (du_1/dx)_{x=-b}. \end{aligned} \tag{4.76}$$

There then arise four simultaneous equations,

$$A + B = C + D \tag{4.77a}$$

$$A\exp[\mathrm{i}(\alpha - k)a] + B\exp[-\mathrm{i}(\alpha + k)a] = C\exp[(\beta - \mathrm{i}k)b] + D\exp[(\beta + \mathrm{i}k)b] \tag{4.77b}$$

$$A\mathrm{i}(\alpha - k) - B\mathrm{i}(\alpha + k) = C(\beta - \mathrm{i}k) - \mathrm{D}(\beta + \mathrm{i}k) \tag{4.77c}$$

$$A\mathrm{i}(\alpha - k)\exp[\mathrm{i}(\alpha - k)a] - B\mathrm{i}(\alpha + k)\exp[-\mathrm{i}(\alpha + k)a] = C(\beta - \mathrm{i}k)\exp[(\beta - \mathrm{i}k)b] - D(\beta + \mathrm{i}k)\exp[(\beta + \mathrm{i}k)b]. \tag{4.77d}$$

For there to be a non-trivial solution, the determinant of the coefficients of A, B, C and D in Equations (4.77) must be equal to zero. This gives, after a lot of manipulation,

$$[(\beta^2 - \alpha^2)/2\alpha\beta] \sinh \beta b \sin \alpha a + \cosh \beta b \cos \alpha a = \cos k(a + b). \tag{4.78}$$

Kronig and Penney made an important simplification which helps us see the physical significance of the result. We let the height of the barrier $V_0 \rightarrow \infty$ and the width of the barrier $b \rightarrow 0$ while at the same time requiring that the area of the barrier $V_0 b$ remains finite. This is mathematically described as a delta function. Under this circumstance, Equation (4.78) reduces to

$$(P/\alpha a) \sin \alpha a + \cos \alpha a = \cos ka, \tag{4.79}$$

where

$$P = meV_0 ba/h^2 \tag{4.80}$$

which is a measure of the barrier area. The left hand side of Equation (4.79) is plotted in Fig. 4.20 for $P = 3\pi/2$, where the abscissa is a measure of energy. Now we note that the right hand side of Equation (4.79) cannot be larger than $+1$ nor less than -1. Thus, only the shaded areas in Fig. 4.20 correspond to real solutions. This is equivalent to saying that only certain values of energy are allowed, or in other words, there are forbidden values of energy. Once again, because of the periodic potential, we have energy bands formed. The width of the allowed band increases with αa, i.e. energy. It is possible to use Equation (4.79) to obtain a plot of E versus k for this model. This is sketched in Fig. 4.21 for $P = 3\pi/2$ and shows all the features illustrated in our earlier discussion of the nearly free electron model.

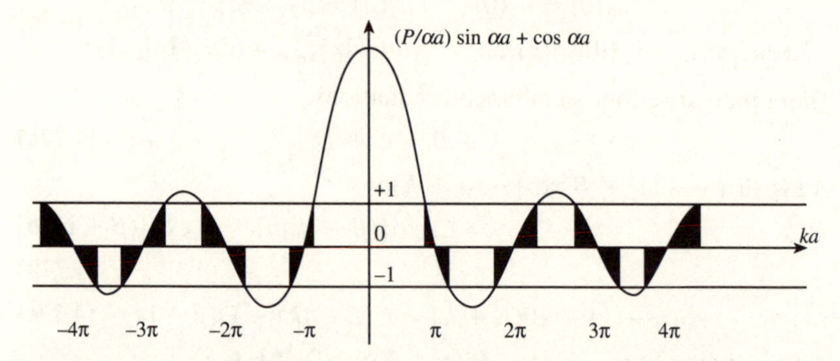

Fig. 4.20 Plot of $(P/\alpha a) \sin \alpha a + \cos \alpha a$ versus ka for $P = 3\pi/2$. The only allowed solutions are for ordinate values between $+1$ and -1.

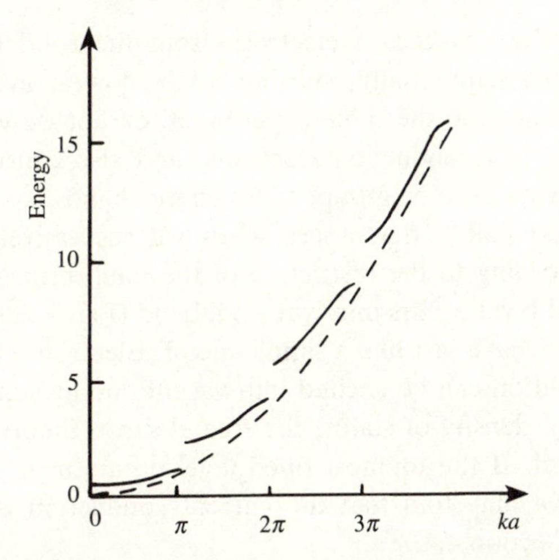

Fig. 4.21 Plot of energy versus wavevector for the Kronig–Penney model with $P = 3\pi/2$. Ordinate in units of $h^2/8ma^2$. The dashed line corresponds to the free electron approximation.

(The result can also be obtained by solving the central equation for a delta function potential $V(x) = (h^2/2me)Ra\delta(x - a)$ where R is a constant, a the lattice spacing and δ is the Kronneker delta function. For such a function all $U_g = R$. This is treated by S. Singh, *Am. J. Phys.* **51** (1983) 179 and is left as an exercise for the student.)

4.7 Metals, semiconductors and insulators

Our discussion of energy bands has now reached a point where we can identify the differences between metals, semiconductors and insulators and then go on to examine which elements we might expect to form solids in these various classes. First let us recall that:

1 Metals have conductivities of the order of 10^5 (ohm cm)$^{-1}$ which decrease with increasing temperature
2 Insulators have conductivities as low as 10^{-28} (ohm cm)$^{-1}$ which rise with temperature
3 Semiconductors have conductivities intermediate between the two which also arise with temperature.

In our discussion of the quantum mechanical free electron model (Section 2.1.1), we performed a 'thought experiment' in which we removed all the *conduction* electrons from the solid and filled the quantum states sequentially, starting at the lowest available energy levels. Let us now do the same experiment, except we will remove all electrons (not just conduction electrons) and also consider the available quantum states to be grouped into energy bands.

The first possibility that arises when we successively replace the electrons according to the restriction of the Pauli principle is that the topmost filled level occurs mid-way up a band (Fig. 4.22(*a*)). Then the material will behave just like a simple metal. Electrons close to the top of the distribution can be excited into vacant energy states and except for a different density of states, the free electron theory of Chapter 2 can be applied. If the topmost filled level occurs near the top of the band, then we may find that the current conduction is by holes, as explained in Section 4.4.2.

An alternative possibility is that the topmost filled level comes at the top of a band (Fig. 4.22(*b*)). Then there are no available states into which the electrons near the top of the distribution can be thermally excited. Conduction cannot take place, and the material is an insulator. Of course, it is not a perfect insulator because there is a small, but non-zero, probability that electrons will be thermally excited into the empty band (known as the conduction band) from the lower, full, band (known as the valence band). This small conductivity is strongly temperature dependent as we shall see in Chapter 6, but we already

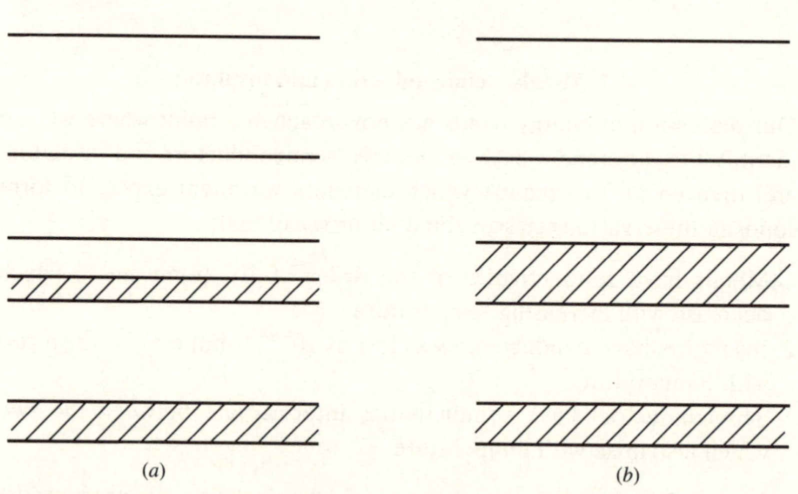

Fig. 4.22 Schematic energy band diagrams of (*a*) a metal with a partially filled conduction band, (*b*) an insulator with a just filled valence band.

see that it will rise with temperature, as more and more electrons are thermally excited across the energy gap.

While we tend to make a distinction between semiconductors and insulators, the distinction is not fundamental. If the band gap is small, then there will be a very significant population of the states in the conduction band at ambient temperatures. Significant electrical conduction occurs. Nevertheless, the pure semiconductor band structure resembles an insulator insofar as it has a filled valence band and empty conduction band. The difference is in the size of the energy gap. As we shall see, there is an important class of semiconductors which can be doped with specific impurities in order to vary their conductivity over a wide range.

From the above discussion, it might seem that most materials should exhibit metallic properties and that insulators and semiconductors should, statistically, be a rarity. After all, you have to get the topmost filled level *exactly* at the top of the band in order to get insulating properties. However, if we go back to the tightly bound electron model and look to see how energy bands form, then from the electron configuration of the atom we should be able to make some deductions about the band structure of the solid.

4.7.1 The alkali metals

Let us begin with a simple case, that of the alkali metals. The electronic structure of the sodium atom for example is, using spectroscopic notation, $1s^2 2s^2 2p^6 3s^1$. This is a shorthand way of denoting the states and number of electrons in those states. The large number gives the shell number, that is the n number of the Bohr theory. Electrons with different n numbers have very different energies. The letters, s, p etc. refer to the subshells within these levels. They refer, in fact to states of different orbital angular momentum. The s level has no orbital angular momentum and therefore can accommodate only two electrons, of opposite spin states. The p level has angular momentum $\hbar$ and quantum theory gives six possible states when we include the spin angular momentum. We denote the number of electrons in these subshells by the superscript figure. Thus $1s^1$ is the configuration of the hydrogen atom, i.e. it has one electron in the s subshell of the $n = 1$ shell. Helium has the configuration $1s^2$, i.e. two electrons in the s subshell of the $n = 1$ shell. There can be no p subshells in the $n = 1$

shell, so our next atom, lithium, has an electronic configuration $1s^22s^1$, two electrons in the $n = 1$ shell and one electron in the s subshell of the $n = 1$ shell. *Lithium* has a full 1s subshell but only a half full 2s subshell. Let us refer to Fig. 4.12 to see how the electronic structure goes over into the band structure of the solid of N atoms. Each discrete atomic level gets split up into N states in the quasi-continuum in the solid. Thus the 1s shell splits up into a band of $2N$ levels well separated from the 2s band, also containing $2N$ levels. Each electron in the 1s shell therefore has a state in the 1s band of the solid. There are thus $2N$ electrons occupying the $2N$ states and the band is exactly full. However, the 2s shell gives rise to $2N$ states in the 2s band but there is only one electron per atom in the 2s state. Only N electrons occupy the $2N$ states in the 2s band and the Fermi level comes exactly halfway up the band. Lithium behaves like a free electron metal because electrons near the Fermi level can be easily excited to vacant states.

Sodium is the next alkali metal in the periodic table, one row down from lithium. As already stated, the atomic configuration is $1s^22s^22p^63s^1$. It is similar to lithium in that the outer subshell is again only half full. The $n = 2$ shell is now full, containing eight electrons. Reference to Fig. 4.12 shows how the electronic structure transforms from the atom to the solid containing N atoms. Again, because each atomic level corresponds to N levels in the solid, the outer 3s band is only half full and the solid has metallic properties.

4.7.2 Noble gas solids

At room temperature, the elements such as neon, argon, krypton and xenon are in the gaseous state and, because of their almost inertness chemically, are known as the noble gases. At low temperatures they do solidify, forming close packed crystal structures. These solids are excellent insulators, as we can understand by examining their electronic structure in the same way that we did for the alkali metals. Let us take neon as the simplest example. *Neon* has the structure $1s^22s^22p^6$ so, unlike sodium, the 3s levels are completely empty. In the solid, the 1s, 2s and 2p bands are therefore full and the 3s band is empty. As the 3s band is separated from the 2p band by a considerable distance, many times k_BT at room temperature, electronic conduction cannot take place easily. Solid neon is therefore an excellent insulator. A similar argument applies for all the noble gas solids.

4.7.3 Alkaline earths

Would that life were simple! Moving along the third row of the periodic table from sodium brings us to magnesium. *Magnesium* has an electronic structure $1s^22s^22p^63s^2$. At first sight we might expect magnesium to be an insulator because the 3s subshell is completely full. Unfortunately, this immediate conclusion is wrong, as magnesium is a normal metal, having quite high conductivity. The answer lies in the overlapping of the 3s and 3p bands in the solid. In one dimension, band overlap is forbidden; the perturbation introduced by neighbours can only broaden the groups of energy levels but in three dimensions overlap is quite possible. The s and p subshells are quite closely spaced in the atom and in the solid the broadening of these levels into bands leads to overlap (Fig. 4.23). Electrons are then transferred from the higher energy states of the 3s band into the lower 3p states and the Fermi level lies part way up the overlapping 3s and 3p bands. Electrons at the Fermi level see vacant states in the quasi-continuum into which they can be excited and metallic properties are observed.

4.7.4 The diamond structure

Carbon has an electronic structure $1s^22s^22p^2$ and we would expect it to exhibit metallic properties. We know, however, that diamond is one of the best insulators! The explanation of this anomaly is that the 2s and 2p states do not retain their own separate character but hybridize. Four

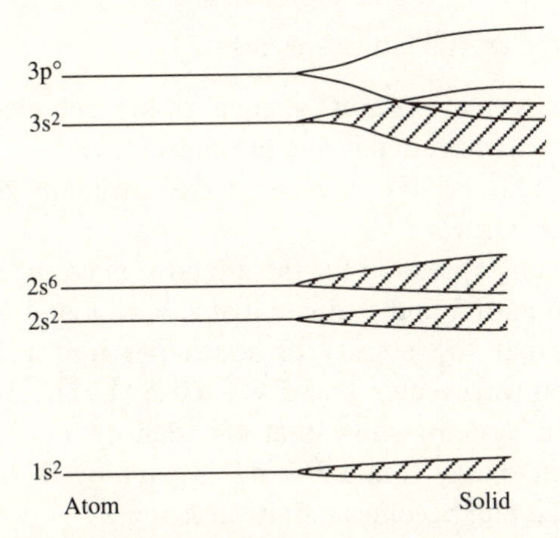

Fig. 4.23 Schematic diagram showing the broadening of atomic levels with increasing interaction and the presence of band overlap, resulting in metallic behaviour in the alkaline earth magnesium.

hybrid states are formed, linear combinations of the atomic orbitals. These four states are equivalent in energy and electron density. They give rise to four very strongly directional bonds which characterize the diamond structure. As we shall see in Chapter 6 these strongly directional bonds are very important in determining the properties of the semiconductors silicon and germanium, both of which have the diamond structure.

Problems

4.1 According to the nearly free electron model, energy gaps occur at values of wavevector corresponding to Bragg reflection. Assuming that the crystal potential makes only a very small perturbation on the free electron potential, calculate the directions in a simple cubic structure in which the first, second and third energy gaps occur. Calculate the electron energies at which these gaps occur for a lattice constant $a = 4 \times 10^{-10}$ m.

4.2 Use the result of the Kronig–Penney model (Equation (4.79)) to determine the energy of the lowest band at wavevector $k = 0$ in terms of V_0, a and b. Comment on this in relation to the free electron model.

4.3 The energy versus wavevector curve for conduction electrons in a hypothetical metal has the form

$$E = 2A \sin^2 (ka/2)$$

where a is the crystal lattice spacing.

(a) If the effective mass m^* is equal to the free electron mass for low wavevector, determine the constant A.

(b) Verify that $\partial E/\partial k$ is zero at the Brillouin zone boundary where $k = \pm\pi/a$.

(c) Derive an expression for the effective mass m^*. Sketch m^* as a function of k and evaluate it at $k = \pi/2a$ and $k = \pi/a$.

(d) Given that the density of states per unit volume $D(k)\mathrm{d}k$ between wavevector k and $k + \mathrm{d}k$ is $(1/\pi^2)k^2\mathrm{d}k$ as for a free electron system, show that the density of states $D(E)\mathrm{d}E$ between energy E and $E + \mathrm{d}E$ is given by $D(E) = m/4\hbar^2 a$ at $k = \pi/2a$ and becomes infinite at $k = \pi/a$.

4.4 The band structure of a certain one-dimensional metal is described by the tightly bound electron approximation. The energy E of electrons in the conduction band is

$$E = E_0 - A\cos(ka)$$

where k is the electron wavevector and E_0 and A are constants. Derive an expression for the density of states $D(E)\mathrm{d}E$ between energy E and $E + \mathrm{d}E$.

If the Fermi energy occurs at a wavevector of $k_F = \pi/3a$, prove that the electronic specific heat capacity is given by

$$C_e = nk_B^2 2\pi T/(aA3\sqrt{3}).$$

4.5 The energy E of an electron in the conduction band of a one-dimensional metal is related to the wavevector k by

$$E = A\sin^2 ka$$

where A is a constant and a is the atomic spacing. Determine the width of the first Brillouin zone.

If $a = 2 \times 10^{-10}$ m, the Fermi wavevector is $2.618 \times 10^9\ \mathrm{m}^{-1}$ and the Fermi energy is 1 eV, determine the effective mass of the conduction electrons measured in a resonance experiment, which probes the electrons at the Fermi level.

4.6 In a particular direction in a crystal, within the first Brillouin zone the energy E of an electron varies with wavevector k as

$$E = Ak^2 + Bk^4.$$

If the effective mass equals the free electron mass m at low values of k, determine the value of the effective mass as a fraction of m, at the first Brillouin zone boundary $k = \pi/a$.

5

Experimental evidence for band structure and effective mass

At the beginning of the previous chapter, we reviewed some experimental data which could not be explained using the free electron model. While the Hall effect unequivocally indicates the breakdown of the free electron model, it does not provide direct evidence for the existence of energy bands. There are, however, a number of techniques which do provide a direct measure of the energy gaps and the density of states.

5.1 Optical techniques for band structure measurements

5.1.1 Infra-red absorption in semiconductors

The first technique is both the easiest to understand and the easiest to perform experimentally. If one looks at a piece of polished silicon or germanium, it has the appearance of a metal. However, if thinned to below a few micrometres in thickness a piece of silicon is translucent, having a red appearance. A certain amount of light is transmitted in the red end of the spectrum. If one goes further into the infra-red, we find that these semiconductors are transparent.

Monochromatic radiation can be obtained from the continuous spectrum of infra-red radiation emitted by a hot filament by use of a diffraction grating spectrometer. The intensity transmitted through the semiconductor is measured as a function of wavelength, and, as illustrated in Fig. 5.1, a very abrupt drop in transmission is observed at a frequency characteristic of the semiconductor. For germanium and other common semiconductors this 'absorption edge', as it is called, occurs at around 1–2 μm wavelength. This is in the near infra-red.

We can understand the origin of this absorption edge by considering how electromagnetic radiation is absorbed in the semiconductor. When

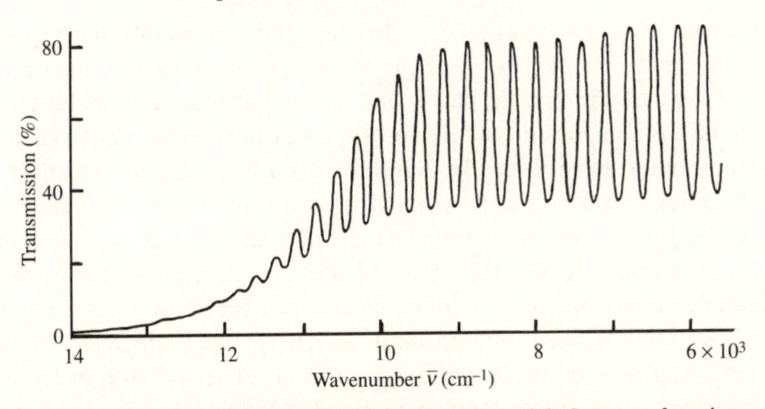

Fig. 5.1 Intensity transmitted through a thin crystal of $SnSe_2$ as a function of wavenumber (=(wavelength)$^{-1}$). The sharp drop is termed the absorption edge, from which the band gap is deduced to be 1.000 ± 0.03 eV. The oscillations arise from interference in the light reflected from the top and bottom surfaces of the thin sample. (After P. A. Lee and G. Said, *Brit. J. Appl. Phys*. **1** (1968) 837.)

this occurs, an electron is excited from the valence band across the energy gap into the conduction band, leaving a hole in the valence band. These free carriers can be moved through the crystal under a bias voltage applied across it and a photocurrent measured. This is how light dependent resistors such as the ORP12 CdS devices operate. If however, the photon energy $h\nu$ is less than the energy gap E_g, no absorption can take place because insufficient energy is available to excite an electron from the valence to the conduction band. An abrupt increase in absorption (decrease in transmission) occurs when

$$h\nu = E_g. \tag{5.1}$$

Here is a very easy way of measuring the energy gap in a semiconductor and indeed it is an experiment readily adaptable for use in an under-graduate teaching laboratory. (We note that there will be some fine structure on the absorption edge because of the presence of excitation levels, but we will discuss this in the next chapter.)

5.1.2 Absorption studies in metals

The above example was one of the simplest experiments in which electromagnetic radiation is used to probe for discontinuities in the density of states. Wherever there is a discontinuity some feature will appear in the absorption spectrum at an energy just sufficient to excite electrons over the energy gap. In the study of metals, absorption studies in the visible region of the spectrum can yield information on the shape of the energy bands and the band gaps. For metals, the situation is complicated by the absorption of the conduction electrons, and often the absorption edge has to be deduced by fairly sophisticated calculations from reflectivity measurements made over a range of wavelengths. In many materials, the band structure is complex and optical measurements yield only values of the critical gap parameters. Complex band structure calculations are then set up on large computers and the solutions iterated until the energy gaps predicted by the theory agree with those observed. It is a somewhat unsatisfactory approach to the understanding of band structures but it is the best method currently available.

5.1.3 X-ray and ultra-violet photoelectron spectroscopy (XPS and UPS)

Photoelectron spectroscopy is an extremely powerful technique for the study of the valence bands of metals and with rapid developments in ultra-high vacuum technology over the past few years, has become very popular. In these experiments ultra-violet ($h\nu \lessapprox 20$ eV) or soft X-radiation ($h\nu \approx 1$ keV) is shone onto a clean surface of the solid under examination and the energy of the photoelectrons ejected when the radiation is absorbed is analysed. The number of photoelectrons $N(E, h\nu)\delta E$ in the energy range E to $E + \delta E$ is measured. The energy distribution curve is given by $N(E, h\nu)$ and from it one can obtain information on the band structure. The photoelectron can be thought of as being produced in three stages. Firstly a valence band electron is excited to an empty state above the Fermi level, then the excited electron moves to the surface of the material and finally escapes into the vacuum. Thus the probability of absorption of an ultra-violet or X-ray photon is proportional to the product of the density of states in the valence band and the density of states in the continuum above the vacuum level. As this latter distribution is a very smooth function, to a good approximation, the energy distribution curve (when corrected for effects of scattering of the photoelectrons in the material before emergence) gives the density of states in the valence band. Figure 5.2(a) shows the upward transitions of the valence band electrons to the continuum states. Each electron has an added energy $h\nu$ and more electrons are excited from the middle of the band than from the band edges. Provided we make the assumption that all transitions from the band are equiprobable, the energy distribution curve is proportional to the valence band density of states. An example of an X-ray photoelectron spectrum is given in Fig. 5.2(*b*). This is of the rare earth element samarium.

The use of a single frequency of exciting radiation can clearly be seen to be a handicap in unscrambling the complex features seen in

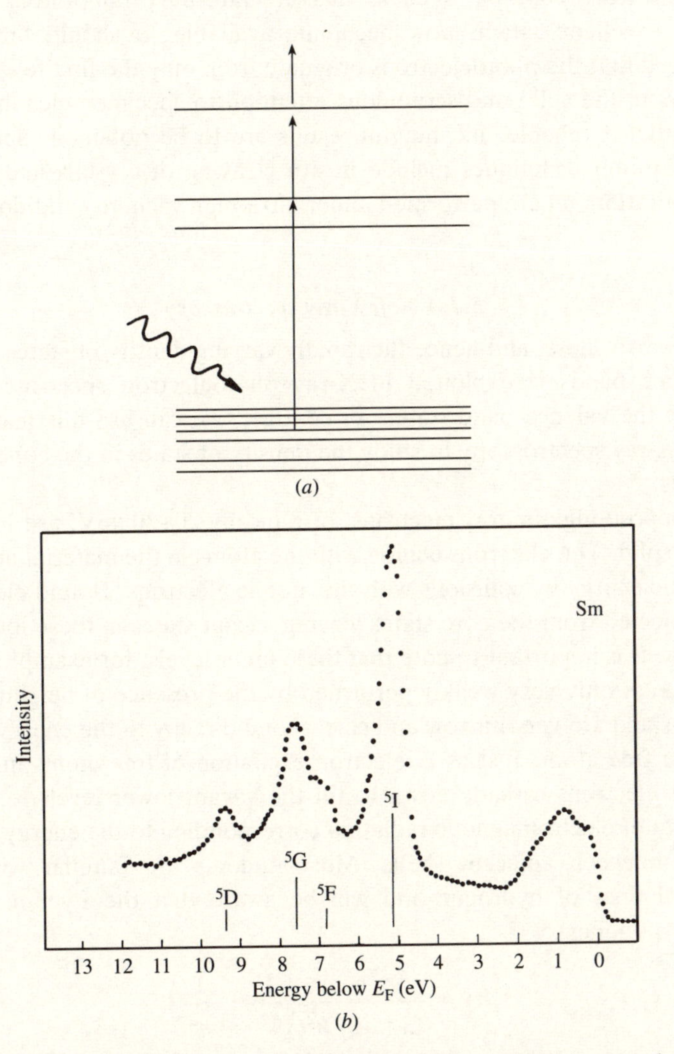

Fig. 5.2 (*a*) Schematic diagram of the removal of a photoelectron from a valence band via the conduction band states. (*b*) X-ray photoelectron spectrum of samarium. (After Y. Bauer and G. Busch, J. Electron Spectroscopy **5** (1974) 611.)

some XPS spectra and in order to deconvolve the effects of changes in the density of the photoelectron final states, photoelectron spectra taken over a range of exciting frequencies are valuable. Until synchrotron radiation became available, tunable frequency sources were not available in the soft X-ray or vacuum ultra-violet regions of the spectrum. However with the establishment of dedicated sources of synchrotron radiation, such as that at Daresbury Laboratory in the UK, excellent data is now becoming available. It should finally be noticed that the photoelectrons originate from only the first few atomic layers in the solid and scrupulous attention to specimen cleanliness is essential if reliable, meaningful results are to be obtained. Specimen preparation techniques include in-situ cleaving of crystals and in-situ evaporation, all are performed under ultra-high vacuum conditions.

5.1.4 Soft X-ray spectroscopy

The narrowness, and hence the rapidly varying density of states, of the valence bands is exploited in X-ray photoelectron spectroscopy to study the valence band states. In contrast, we can use this feature in soft X-ray spectroscopy to study the density of states in the conduction bands.

Suppose high energy electrons, of typically 20–30 keV, are incident on a solid. The electrons collide with the atoms in the material and lose kinetic energy by collisions with the atomic electrons. Bound electrons are ejected from the core states leaving vacant states in these low lying levels. It is important to note that these inner levels, for example the 1s level, are only very weakly perturbed by the presence of neighbouring atoms and are very narrow and correspond directly to the energy levels in the free atom. Just as in electron excitation of free atoms in a gas, outer electrons cascade down to fill the vacant lower levels, emitting quanta of electromagnetic radiation corresponding to the energy difference between adjacent shells. Most students are familiar with the spatial lines of hydrogen and will be aware that the Lyman series, corresponding to

$$h\nu = \frac{Z^2 e^4 m}{2(4\pi\varepsilon_0)^2\hbar^2}\left(\frac{1}{1^2} - \frac{1}{n^2}\right) \tag{5.2}$$

where n is an integer and $Z = 1$ for hydrogen, is in the near ultra-violet region of the spectrum.

While the simple Bohr model cannot be applied to the more

complex spectra of heavier elements, Equation (5.2) does show that the frequency of the radiation emitted during a transition from the $n = 2$ to $n = 1$ shell is strongly dependent on the atomic number Z. Indeed by the time one reaches copper ($Z = 29$) the energy of this spectral line is 8 keV, corresponding to a wavelength of 1.54 Å. These are quite hard X-rays and this characteristic emission line, which is still quite sharp because of the screening of these core electrons from the effects of neighbouring atoms, is a prominent feature of the spectrum from an X-ray generator. Indeed, so strong and sharp is this line that the copper K line, as it is known, is used extensively as a monochromatic X-ray source for X-ray crystallography. In some areas of crystallography its use is almost universal.

The hard X-ray characteristic lines are sharp because of the sharpness of the inner core levels, but the outer bands are, as we have seen, very much broadened by the perturbation of neighbouring atoms. Thus if we can study the X-ray emission associated with an electronic transition from an outer band to an inner band, then the distribution of intensity reveals the density of states in the conduction band.

Let us take sodium as a simple case where only the conduction band is appreciably broadened. Transitions from the 2p to the 1s band will give sharp lines but transitions from the 3s band to the 2p band will give a broad spatial band (Fig. 5.3). The intensity at frequency ν is dependent on the density of states in the 3s band, as the 2p band is effectively a single energy. Microdensitometry or electronic recording

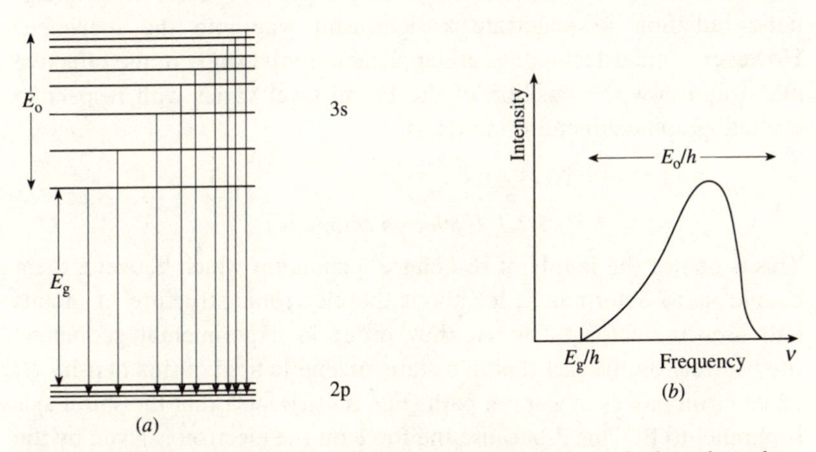

Fig. 5.3 (*a*) Schematic diagram showing the transitions from the broad conduction band to narrow valence band. (*b*) X-ray emission spectrum corresponding to the transitions in (*a*).

of the spectrum gives the density of states, provided one again makes the assumption that all transitions are equiprobable.

As can be seen from Fig. 5.3, the soft X-ray spectrum gives a number of parameters. From the intensity, we obtain the density of states while from the minimum frequency of the spectral band we obtain E_g/h. The frequency width of the spatial band gives the energy width of the conduction band. However, it is worth re-iterating that, as for photo-electron spectroscopy, ultra-high vacuum conditions ($\leqslant 10^{-10}$ torr) are essential if meaningful results are to be obtained. Due to the high absorption of soft X-rays in air, the whole experiment must be under vacuum. In-situ evaporation of the metal under investigation onto a clean substrate is also really essential. Under normal high vacuum pressure of 10^{-6} torr, in a few minutes the whole surface will be covered with one or two layers of adsorbed molecules and in these experiments which are so surface sensitive, clean conditions are prerequisite. This unfortunately not only puts up the cost of performing the experiments but greatly adds to the complexity of carrying them out.

5.2 Measurement of effective mass

In this section we will concentrate primarily on techniques for the measurement of the effective mass of electrons in semiconductors. This is because semiconductors show wide variations in effective mass and also because the low carrier concentrations present enable electromagnetic radiation to penetrate a significant way into the specimen. However, similar techniques are applicable to the study of the effective mass (and how the position of the Fermi level varies with respect to crystallographic direction) in metals.

5.2.1 Cyclotron resonance

This is one of the family of resonance techniques which between them enable us to determine a lot about the electronic structure of metals and semiconductors. Though they differ in experimental geometry, they all rely on the fact that in a static magnetic field of flux density B, an electron moves in a spiral path (Fig. 5.4(a)) such that the spiral axis is parallel to **B**. This is because the force on the electron is given by the Lorenz force

$$\mathbf{F} = e\boldsymbol{v} \times \mathbf{B}, \tag{5.3}$$

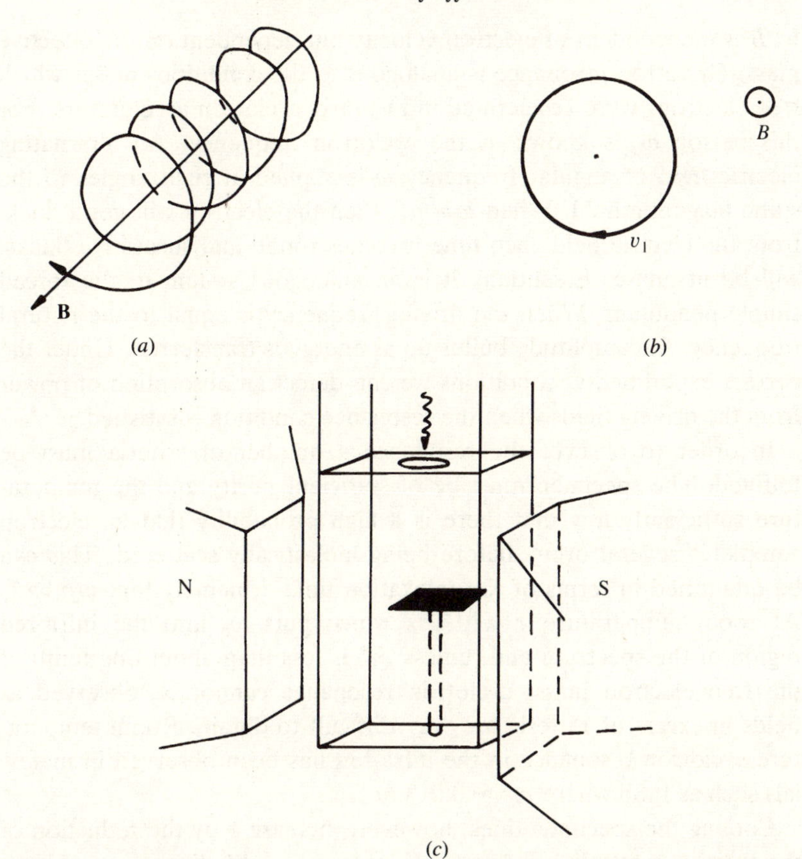

Fig. 5.4 (*a*) Spiral path of an electron in a static magnetic field. (*b*) Projected trajectory normal to the field direction. (*c*) Schematic experimental arrangement for cyclotron resonance at microwave frequencies.

where $\boldsymbol{v}$ is the velocity of the electron, and this force is always directed normal to the velocity vector. If we look at the projection of the electron motion in a plane perpendicular to **B**, we see the projected electron motion is a circle radius r (Fig. 5.4(*b*)). Let us denote the component of velocity in this plane v_1. Then the force towards the centre of the circle is ev_1B and thus

$$m^*v_1/r = ev_1B. \tag{5.4}$$

Writing $\omega_c = v_1/r$ we have

$$\omega_c = eB/m^*. \tag{5.5}$$

In other words, the orbit angular frequency ω_c in a plane perpendicular to B is independent of electron velocity but dependent on the effective mass. Cyclotron resonance is analogous to the conditions under which free electrons were accelerated in the early cyclotron accelerators. For this reason ω_c is known as the cyclotron frequency. An alternating electric field of angular frequency ω is applied at right angles to the static magnetic field. When $\omega = \omega_c$, then the electron will get a 'kick' from the electric field each time it comes round and successive 'kicks' will be in phase. Essentially it is an analogous system to the forced simple pendulum. When the driving frequency is equal to the natural frequency, the amplitude builds up as energy is transferred. Under the correct experimental conditions we can detect an absorption of power from the driving fields when the resonance condition is satisfied.

In order to observe the resonance a number of criteria must be fulfilled. The specimen must be of sufficient purity and the temperature sufficiently low that there is a high probability that an electron completes several orbits before being inelastically scattered. This can be quantified in terms of the relaxation time τ, namely that $\omega_c\tau \gg 1$. At room temperature $\tau \approx 10^{-12}$ s which puts ω_c into the infra-red region of the spectrum and, unless m^* is less than about one tenth of the free electron mass, cyclotron resonance cannot be observed as fields in excess of 15 tesla are very difficult to obtain. Room temperature cyclotron resonance in the infra-red has been observed in materials such as InSb where $m^* \approx 0.013\ m$

Cooling the specimen does, however, increase τ by the reduction of the thermal scattering and at 4.2 K, that is liquid helium temperature, work at microwave frequencies is possible. For 3 cm microwaves (X band) $\omega \approx 60$ GHz, if $m^* = m$ we have $B \approx 0.4$ tesla, a field which is very easy to obtain in the laboratory. As sketched in Fig. 5.4(*c*), the sample is mounted in a half wavelength microwave cavity in such a position as to be at the maximum of the microwave electric field. The frequency of the microwaves is difficult to vary appreciably so the magnetic field is swept until absorption of microwave power is detected. One problem about working with semiconductors at low temperatures is that very few free electrons are present as thermal excitation across the band gap is not very probable. In order to create sufficient carriers to detect the cyclotron resonance absorption, free carriers are created by shining light down the waveguide, thereby exciting electrons across the band gap into the conduction band. Table 5.1 shows representative values of the effective masses of electrons in some common semiconductors.

Table 5.1 *Effective masses of electrons in some typical semiconductors*

Material	m^*/m
Si	0.98
Ge	1.57
InSb	0.013
InP	0.073
GaAs	0.07

It is, however, very important to emphasize that due to the complexities of the band structures in real semiconductors there is more than one value for the effective mass and which value is observed depends on the experiment and its geometry. These questions cannot be discussed without a rather more detailed study of band structure than is possible here.

5.2.2 Plasma resonance

A number of optical techniques exist which enable one to obtain a parameter dependent on the free carrier concentration and the effective mass. If the carrier concentration is known from Hall effect measurements, m^* can be deduced. The example chosen here is known as plasma resonance. We are all familiar with the characteristic lustre of metals in visible light, but many of us are not aware that in the ultra-violet the alkali metals become transparent. We can understand the origin of this absorption edge, found in some semiconductors in the far infra-red, from a classical theory of the interaction of electromagnetic radiation with a gas of free electrons.

Maxwell's equations of electromagnetism can be combined to give a wave equation describing the electric field vector **E** of the light wave. If the light propagates along the x axis, this equation is

$$\frac{\mathrm{d}^2\mathbf{E}}{\mathrm{d}x^2} = \frac{\mu\varepsilon}{c^2}\frac{\mathrm{d}^2\mathbf{E}}{\mathrm{d}t^2} \tag{5.6}$$

where μ is the relative permeability, ε the dielectric constant of the medium and c is the velocity of light in vacuo. We should note that the direction of the electric field vector is transverse to the propagation direction, say in the y direction. Let us suppose the material is non-magnetic, i.e. $\mu = 1$. Then if we look for a travelling wave solution of the form $\exp[i(kx + \omega t)]$ we see that

$$k^2 = \varepsilon\omega^2/c^2. \tag{5.7}$$

Now the velocity of light in the medium v is

$$v = \omega/k \tag{5.8}$$

and the refractive index n is then

$$n = c/v = ck/\omega. \tag{5.9}$$

Combining Equations (5.7) and (5.9), we have

$$\varepsilon = n^2 \tag{5.10}$$

at optical frequencies.

We must now determine the dielectric constant of the free electron gas. To an electron, the electric field of the incident light is purely oscillatory, of the form

$$E = E_0 \exp(i\omega t). \tag{5.11}$$

The equation of motion of the electron is then

$$m^* \frac{d^2y}{dt^2} + \frac{m^*}{\tau}\frac{dy}{dt} = eE_0 \exp(i\omega t) \tag{5.12}$$

where τ is again the relaxation time and this has a solution,

$$y = \frac{-ieE_0\tau e^{i\omega t}}{m^*\omega(1 + i\omega\tau)}. \tag{5.13}$$

Such a displacement of the electron is equivalent to the production of an oscillating electric dipole moment. For N such electrons the induced dipole moment (per unit volume) P is given by

$$P = \frac{-iNe^2E_0\tau e^{i\omega t}}{m^*\omega(1 + i\omega\tau)}. \tag{5.14}$$

This induced dipole moment per unit volume is known as the polarization. The electric displacement D is defined in terms of the polarization P and electric field E, namely

$$D = \varepsilon_0 E + P, \tag{5.15}$$

and as the dielectric constant ε is defined as

$$D = \varepsilon\varepsilon_0 E \tag{5.16}$$

we have

$$P = \varepsilon_0(\varepsilon - 1)\,E. \tag{5.17}$$

Thus

$$\varepsilon_0(\varepsilon - 1) = \frac{-Ne^2\tau}{m^*\omega(1 + \mathrm{i}\omega\tau)}. \tag{5.18}$$

Rearranging, we have

$$\varepsilon = 1 + \frac{\mathrm{N}e^2\tau}{\mathrm{i}m^*\omega\varepsilon_0(1 + \mathrm{i}\omega\tau)}. \tag{5.19}$$

This can be written in terms of its real and imaginary parts by multiplying the top and bottom of the second term on the right hand side by $(1 - \mathrm{i}\omega\tau)$. This gives

$$\varepsilon = \varepsilon_1 - \mathrm{i}\varepsilon_2 = 1 - \frac{Ne^2\tau^2}{m^*\varepsilon_0(1 + \omega^2\tau^2)} - \mathrm{i}\frac{Ne^2\tau}{m^*\omega\varepsilon_0(1 + \omega^2\tau^2)}, \tag{5.20}$$

i.e.

$$\varepsilon_1 = 1 - \frac{Ne^2\tau^2}{m^*\varepsilon_0(1 + \omega^2\tau^2)}. \tag{5.21}$$

At optical frequencies $\omega^2\tau^2 \gg 1$ and so

$$\varepsilon_1 = 1 - \frac{Ne^2}{m^*\varepsilon_0\omega^2}. \tag{5.22}$$

Now if $\varepsilon_1 > 0$, the real part of the refractive index is positive and the wave can propagate through the medium. If, however $\varepsilon_1 < 0$, the refractive index is purely imaginary and this is equivalent to the wave being exponentially damped as it travels into the medium. Light cannot propagate if $\varepsilon_1 < 0$. Thus if

$$\omega^2 < Ne^2/m^*\varepsilon_0 \tag{5.23}$$

no transmission occurs. At

$$\omega = \omega_\mathrm{p} = (Ne^2/m^*\varepsilon_0)^{\frac{1}{2}} \tag{5.24}$$

a sharp change in transmission occurs and this is known as the plasma edge. From a measure of ω_p, known as the plasma frequency we

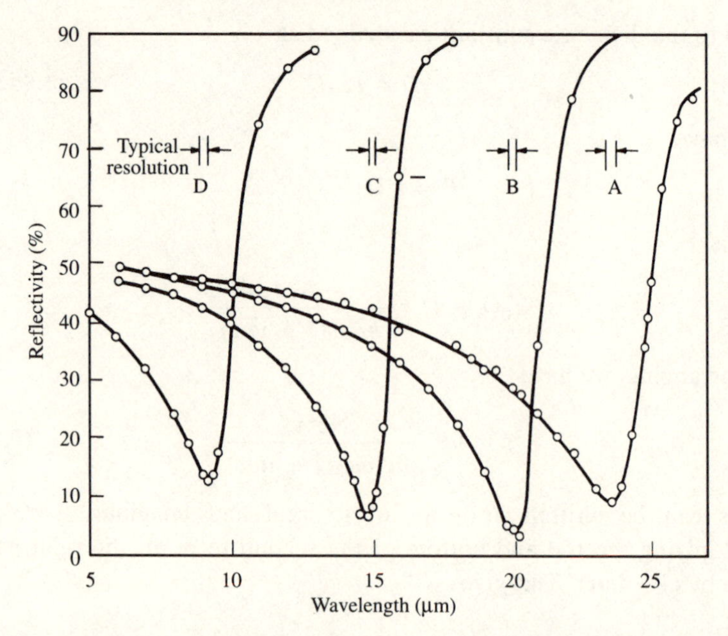

Fig. 5.5 Example of a plasma resonance edge in p-type PbTe. Reflectivity rises to near 100% below the edge. Curve A: 3.5×10^{18} cm^{-3}, Curve B: 5.7×10^{18} cm^{-3}, Curve C: 1.5×10^{19} cm^{-3}, Curve D: 4.8×10^{19} cm^{-3} hole concentration. (After J. R. Dixon and H. R. Reidl, *Phys. Rev.* **138** (1965) 873.)

obtain N/m^* and if N is known, m^* can be deduced. Figure 5.5 shows typical plasma edges in lead telluride with various hole concentrations. For semiconductors with low carrier concentration, ω can be in the far infra-red.

6

Electrical conduction in semiconductors and insulators

6.1 The diamond structure

Diamond is an extremely good insulator and one of its remarkable properties is the four, very strongly directional, covalent bonds which hold the crystal together. The bonds are all equivalent and thus symmetry dictates that they should make equal angles with one another. Figure 6.1(*a*) shows this arrangement, the bonds having an angle of 109° with each other. When linked to neighbouring atoms, we see that each atom is surrounded by four others (Fig. 6.1(*b*)), on the apices of a tetrahedron in space. While not immediately apparent, this tetrahedral bonding gives rise to a face centred cubic lattice structure (Fig. 6.1(*c*)) known as the 'diamond cubic' structure.

The details of the crystallography are not of direct concern to us here. For the present purposes it suffices to note that these covalent bonds are very strongly directional. This has important implications

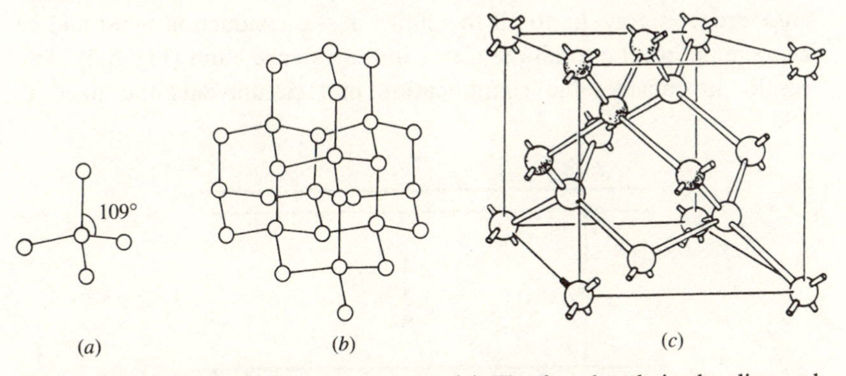

Fig. 6.1 The silicon (diamond) structure. (*a*) The four bonds in the diamond structure. (*b*) Tetrahedral bonding of adjacent atoms. (*c*) The resulting face centred cubic lattice.

when we consider the effects of impurities on the electrical properties of the technically important semiconductors silicon and germanium, both of which have the diamond cubic crystal structure.

6.1.1 The narrow band model

As we saw in Chapter 4, the pure semiconductor has a filled valence band and empty conduction band. It differs from an insulator only in the band gap and, indeed, the distinction is not clear-cut between the two classes of materials. Clearly, at absolute zero of temperature, a pure semiconductor (also known as an 'intrinsic' semiconductor) will be a perfect insulator as no electrons will be present in the conduction band and the valence band will be full. Electrical conductivity occurs by excitation of electrons in the valence band across the band gap. As the number of current carriers therefore varies very rapidly with temperature, in contrast to the case for metals, we find the conductivity rising very rapidly with temperature. To determine the dependence of the conductivity on temperature, we must know the density of states in the valence and conduction bands. Initially we will use a very simple model for the band structure and extend the results to a density of states which is more representative of that in a real semiconductor. The model chosen is known as the *narrow band model* and in it we assume that the widths of the energy bands are narrow compared with the energy gap between them. Then we can make an important approximation that all states in the band have the average energy of those states in the band. (Note that this is a similar type of assumption to that made in the Drude theory in Chapter 1.) We can thus ascribe an average energy E_c to all the states in the conduction band and an average energy E_v to all the states in the valence band (Fig. 6.2). This results in considerable simplification and circumvents the need to

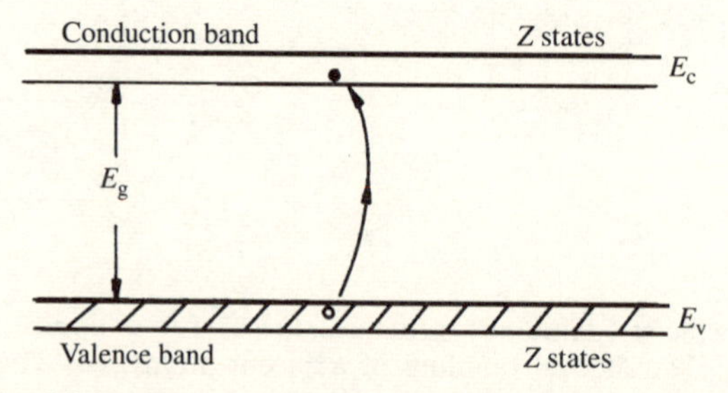

Fig. 6.2 Band structure under the narrow band model.

integrate over the density of states. In my opinion it aids physical insight by removing mathematical complexity.

6.1.2 Conductivity of intrinsic semiconductors

Suppose that an intrinsic (pure) semiconductor has a narrow conduction band, average energy E_c, with Z states in the band and a narrow valence band, average energy E_v, also with Z states in the band (Fig. 6.2). The number of electrons n_c in the conduction band is given, quite generally, by the product of the number of states Z and the probability of occupation $P(E)$, where $P(E)$ is the Fermi–Dirac distribution. Thus,

$$n_c = \frac{Z}{\exp[(E_c - E_F)/k_B T] + 1} \tag{6.1}$$

where E_F is the Fermi energy.

We have, however, run into an immediate problem. What is E_F? According to our previous ideas, E_F marked the energy of the topmost filled level of a metal at absolute zero.

Many students ask 'How can the Fermi energy lie in the forbidden gap, when there are no available levels in the gap?' As has been emphasised before, the Fermi–Dirac distribution is not dependent in any way on the density of states. It arises entirely from statistical mechanical considerations and E_F is simply a *normalizing parameter* in the distribution. We can define it, quite generally, as the energy at which the probability of occupation is $\frac{1}{2}$. That is, to re-quote Equation (2.20),

$$P(E_F) = \tfrac{1}{2}. \tag{6.2}$$

A little thought will reveal that E_F cannot therefore correspond to the topmost level in the valence band. At non-zero temperature, electrons will be excited across the gap and because each such excited electron must leave a hole behind in the valence band there is a symmetry across the gap. If the probability of occupation is then still $\frac{1}{2}$ at the top of the valence band, this symmetry is impossible unless the probability of occupation is also $\frac{1}{2}$ at the bottom of the conduction band. There are then two competing positions for E_F, and clearly E_F can neither lie at the top of the valence band nor at the bottom of the conduction band. It lies in between, in the forbidden gap.

Let us return to the main derivation. Equation (6.1) now gives us n_c but in terms of an undetermined E_F, which we have decided lies in the energy gap. Nevertheless, we can determine its value. The number of electrons in the valence band n_v is similarly given by

$$n_v = \frac{Z}{\exp[(E_v - E_F)/k_B T] + 1} \qquad (6.3)$$

and the number of holes in the valence band p_v is therefore

$$p_v = Z - n_v. \qquad (6.4)$$

Now the number of electrons in the conduction band n_c must exactly equal the number of holes in the valence band p_v because all electrons in the conduction band must have been excited across the gap. Thus, from (6.1), (6.3) and (6.4) we have,

$$1 - \frac{1}{\exp[(E_v - E_F)/k_B T] + 1} = \frac{1}{\exp[(E_c - E_F)/k_B T] + 1}. \qquad (6.5)$$

Therefore

$$\exp[(E_c + E_v - 2E_F)/k_B T] = 1 \qquad (6.6)$$

or

$$E_F = (E_c + E_v)/2. \qquad (6.7)$$

The Fermi energy lies mid-way between the conduction and valence bands. Using this result in Equation (6.1) yields

$$n_c = \frac{Z}{\exp(E_g/2k_B T) + 1} \qquad (6.8)$$

where $E_g = E_c - E_v$.

If $E_g \gg 2k_B T$ then the exponential is large compared with unity and Equation (6.8) approximates to

$$n_c = Z \exp(-E_g/2k_B T). \qquad (6.9)$$

The conductivity σ was defined in Equation (1.8) in terms of the current density J defined in Equation (1.6). In terms of the drift velocity $\langle v \rangle$ and electric field $\mathcal{E}$ we have

$$\sigma = ne\langle v \rangle/\mathcal{E} \qquad (6.10)$$

where n is the number of current carriers per unit volume. The mobility μ is defined as

$$\mu = \langle v \rangle/\mathcal{E} \qquad (6.11)$$

and represents the ease with which carriers transport charge. Usually the mobility of holes is much less than that of electrons. Using Equation (6.11) we see that the conductivity of an intrinsic semiconductor is given by

$$\sigma = (n_c\mu_e + p_v\mu_h)e \tag{6.12}$$

where μ_e and μ_h are the mobilities of electrons and holes respectively. Thus,

$$\sigma = Z(\mu_e + \mu_h)e \exp(-E_g/2k_BT). \tag{6.13}$$

6.2 Extrinsic semiconductors

6.2.1 Effects of impurities

If we evaluate Equation (6.13) for typical semiconductors such as silicon or germanium, we find rather low conductivities typically a few $(\text{ohm m})^{-1}$. This is very much lower than that commonly measured and the discrepancy arises because the presence of small quantities of impurity can make a dramatic difference to the conductivity. It is, in fact, the deliberate addition of impurities which enables us to tailor a specific carrier concentration and hence fabricate the wide range of devices presently on the market. The most important of the impurities are those in Groups III and V of the periodic table.

Let us consider the substitution of one silicon atom in the crystal lattice by an atom of a Group V element, for example phosphorus. Phosphorus is pentavalent, i.e. it has five electrons available to form chemical bonds. When it replaces a silicon atom in a silicon lattice the strong directionality of the silicon bonds forces the phosphorus atom to conform to the prevailing tetrahedral bonding and thus only four of the available electrons on the phosphorus form covalent bonds. The fifth cannot bond covalently and is left attached weakly to the phosphorus atom (Fig. 6.3(*a*)). Its importance lies in the fact that the weak

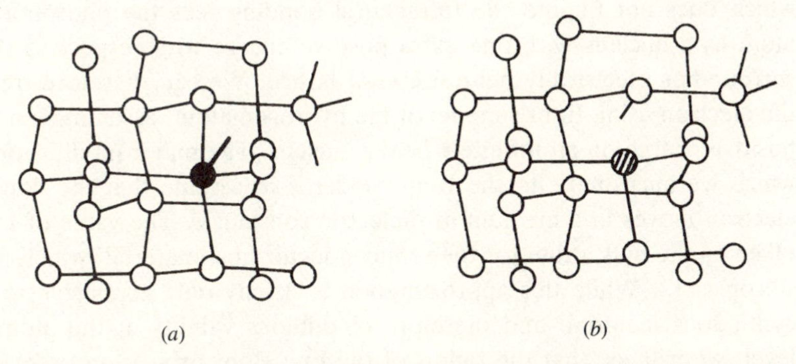

Fig. 6.3 (*a*) Pentavalent impurity in the diamond structure indicating a weakly bound, unpaired electron. (*b*) Trivalent impurity showing the broken bond which gives rise to a hole.

attraction to the phosphorus atom can easily be overcome by thermal excitation and this free electron, now in the conduction band can move through the crystal carrying charge. We call this weakly bound electron a donor electron.

When a trivalent impurity atom from Group III of the periodic table is introduced, a somewhat similar effect arises, except that here it is a weakly bound hole that can be excited into the valence band. Referring to Fig. 6.3(*b*), we see that a trivalent atom, such as boron, is forced to accept the tetrahedral bonding but because there are only three electrons capable of forming covalent bonds, a broken bond exists on one of the neighbouring silicon atoms. Now there are four equivalent silicon neighbours at the apices of the tetrahedron and there is no reason for the broken bond to be attached to any of them, and it is thus shared amongst all four. In the context of the whole crystal this corresponds to a positive charge attached to the impurity atom, i.e. a bound hole. This is called an acceptor state.

6.2.2 Quasi-Bohr model for donors and acceptors

The binding energy of the electrons and holes can be calculated from a very simple model based on Bohr's model of the hydrogen atom. In the case of a pentavalent impurity such as phosphorus, there are only five protons in the nucleus which are of concern, as all the others are effectively screened by the core electrons in inner orbits. From the position of the electrons which form the chemical bonds and are far from the core, the phosphorus atom has one more proton in the nucleus than the surrounding silicon atoms. The weakly bound electron which does not fit into the tetrahedral bonding sees the phosphorus atom as a nucleus with one extra positive charge with respect to the surrounding electrically neutral crystal lattice. We can therefore treat the electron using Bohr's model of the hydrogen atom, assuming a unit positive charge on an infinitely heavy nucleus. The major modification which we must make to the Bohr model is to assume that the donor electron moves in a medium of dielectric constant ε. The value of ε is taken as the bulk value for the semiconductor host material, which for silicon is 12. While this approximation is strictly only applicable to a continuous material, and therefore of dubious validity at the atomic level, we will see that the radius of the first Bohr orbit is an order of magnitude higher than that of the hydrogen. The volume of the first orbit encompasses at least a dozen atoms and we find that the predictions of the model are, if not exact, in the correct order of magnitude. For an electron, charge e, effective mass m^*, moving in a continuum of dielectric constant ε under the influence of a central Coulomb force we have

$$m^* r\omega^2 = e^2/4\pi\varepsilon\varepsilon_0 r^2 \tag{6.14}$$

where ω is the angular frequency, r is the radius of the orbit and ε_0 is the permittivity of free space. According to Bohr's postulate, angular momentum is required in units of $\hbar$, i.e.

$$m^* r^2\omega = n\hbar. \tag{6.15}$$

Thus from Equations (6.14) and (6.15) we have

$$r = 4\pi\varepsilon\varepsilon_0 n^2\hbar^2/e^2 m^*. \tag{6.16}$$

The Bohr orbit radii are larger than those of the hydrogen atom by a factor of $\varepsilon m/m^*$. For silicon the radius of the first Bohr orbit is then 0.64 nm if m^* is equal to the free electron mass. InSb has a very low effective mass, of about 1/100th of the free electron mass and the equivalent radius is over 60 nm!

The kinetic energy T is given by

$$\begin{aligned} T &= m^* r^2\omega^2/2 \\ &= e^4 m^*/2(4\pi\varepsilon\varepsilon_0)^2 n^2\hbar^2 \end{aligned} \tag{6.17}$$

while the potential energy U is given by

$$\begin{aligned} U &= -e^2/4\pi\varepsilon\varepsilon_0 r \\ &= -e^4 m^*/(4\pi\varepsilon\varepsilon_0)^2 n^2\hbar^2. \end{aligned} \tag{6.18}$$

Thus the total energy E_n is given by

$$\begin{aligned} E_n &= U + T \\ &= -e^4 m^*/2(4\pi\varepsilon\varepsilon_0)^2 n^2\hbar^2. \end{aligned} \tag{6.19}$$

The donor electron is in the conduction band when the impurity atom is ionized, i.e. when $n \to \infty$. Therefore we see that the donor electron can occupy a series of localized bound states just below the conduction band. These states are localized because the bound electron is attached to a specific impurity atom and thus these states are represented by short lines in the energy band diagram (Fig. 6.4). The lowest bound state is the ground state of the neutral donor state, i.e. when $n = 1$. This gives the ionization energy (or binding energy) E_b by the difference between the $n = 1$ and $n \to \infty$ state, namely

$$E_b = -e^4 m^*/2(4\pi\varepsilon\varepsilon_0\hbar)^2. \tag{6.20}$$

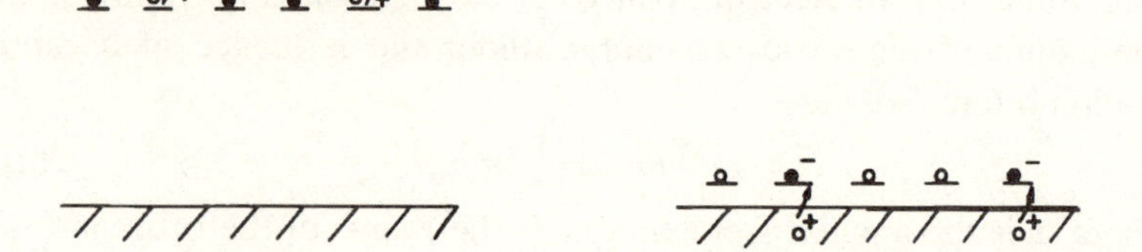

Fig. 6.4 (*a*) Donor and (*b*) acceptor states represented in the energy band diagram.

For silicon $\varepsilon = 12$ and if $m^* = m$ we have

$$E_{\mathrm{b}} = -0.09 \text{ eV}. \tag{6.21}$$

Exactly the same calculation can be applied to the acceptor state with an orbiting hole. The binding energy is then

$$E_{\mathrm{b}} = +e^4 m^*/2(4\pi\varepsilon\varepsilon_0\hbar)^2, \tag{6.22}$$

leading to a set of localized energy levels just above the valence band. Owing to the small binding energies, of only a few times $k_{\mathrm{B}}T$ at room temperature, these donor and acceptor states are known as shallow states. They are by far the most important type of impurity state from a technological point of view.

6.2.3 Thermoluminescence

In many insulators there exist impurity levels which have a ground state well away from the edge of the conduction band. These are known as deep levels and their presence leads to a phenomenon known as thermoluminescence, that is, the emission of light when certain materials are heated to a few hundred degrees centigrade. While it has only recently become a widespread interest, thermoluminescence was discovered three centuries ago by Boyle in 1693. There is a delightful paper in the Royal Society Proceedings in which he describes how he discovered natural thermoluminescence in a large diamond. He tells of how he saw the thermoluminescence glow when he took the diamond to bed with him and it was in contact with his 'naked body'. (We don't write papers quite so spicily these days!)

Thermoluminescence arises from the thermal de-excitation of electrons trapped in deep impurity levels. The presence of impurities can often lead to localized empty states within the forbidden energy region.

If the material is exposed to ionizing radiation, for example cosmic rays, many electron–hole pairs are created as the particle passes through (Fig. 6.5(*a*)). These electrons and holes will gradually 'thermalize', i.e. they will lose energy by collisions with the lattice (exciting phonons), and have energies in the regions close to the band edges. There is a strong probability that the electrons and holes will recombine, and most do so giving off the excess energy (approximately E_g) in the form of light or heat. However, a small fraction of the electrons will be trapped in the deep-lying impurity states (Fig. 6.5(*b*)). Once trapped in these localized potential wells, they can only be returned to the conduction band by thermal excitation and if the level lies more than several k_BT from the band edge, this process becomes extremely unlikely. The probability of an electron having energy E is given by the Fermi–Dirac distribution $\{\exp[(E - E_F)/k_BT] + 1\}^{-1}$ and provided that both conduction band edge and impurity levels lie more than approximately $2k_BT$ above the Fermi level E_F, we can approximate this to $\exp[(E_F - E)/k_BT]$. It immediately follows that the ratio of the probabilities of having an energy corresponding to the conduction band edge E_c and the impurity levels E_I is $\exp[(E_I - E_c)/k_BT] = \exp(-E_b/k_BT)$ where E_b is the binding energy of the electron in the impurity state. For $E_b/k_BT \geqslant 10$ this becomes a very small number, and the probability of de-excitation of the impurity level known as a 'trap' becomes very small unless the sample is raised to a high temperature. If the sample remains at ambient temperature, and is continually exposed to a constant flux of ionizing radiation, whether it be cosmic radiation or the natural

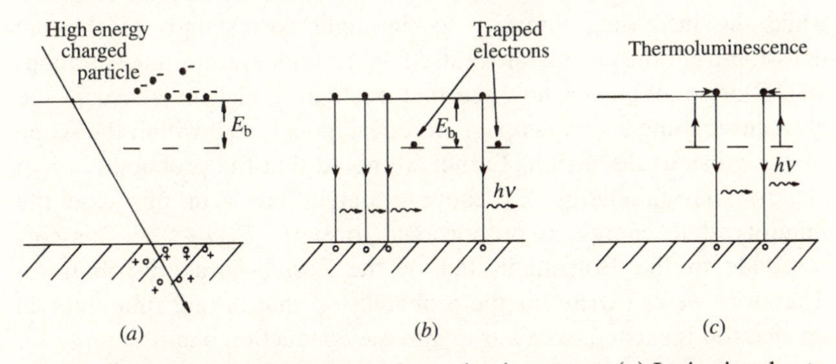

Fig. 6.5 The processes giving rise to thermoluminescence. (*a*) Ionization due to natural or induced radioactivity. (*b*) The trapping of some electrons at deep-level impurity sites. (*c*) De-excitation of the traps on heating, with consequent emission of light.

radioactive emission from rocks, the number of trapped electrons increases linearly with time. Even over very long periods of time the number of trapped electrons acts as a very good measure of the time the material has been exposed to radiation.

Thermoluminescence occurs when material containing electrons trapped in deep impurity states is heated and the electrons thermally excited from the traps. As the temperature rises, the probability of electrons being excited into the conduction band increases dramatically. Once into the conduction band (Fig. 6.5(*c*)) there is only a very small chance that the electrons will be trapped again, the most probable sequence is that the electrons recombine with holes, giving out light. This light is called thermoluminescence, and the amount of light emitted is a measure of the time the object was exposed to radiation. It is now extensively used to date archaeological artefacts, in particular pottery and brick, both containing quartz which is a good thermoluminescent material. There are subtleties involved with dating procedures because the flux of ionizing radiation is often not known, but the basic principle is that indicated. Alternatively, if the time of exposure is known, the amount of light can be used to measure the accumulated radiation dose and this forms the basis of a standard form of radiation monitor. LiF crystals are contained in a small sachet which is incorporated into a badge worn by the radiation worker. Every month the thermoluminescence is measured and the dose received deduced.

The thermoluminescence (or TL) is emitted as the sample temperature is raised and the intensity versus time plot is known as a 'glow curve'. An example is shown in Fig. 6.6. The peaks result from TL while the increasing intensity to the right corresponds to thermal emission from the crystal lattice itself as the temperature reaches about 400 °C or so. We shall now see that one can model the shape of the glow curve using a very simple theoretical model, well within the scope of our previous discussion. Earlier we noted that the probability of an electron having energy E_b above a datum level, in this case the impurity state energy, is proportional to $\exp(-E_b/k_B T)$. (This corresponds to the Boltzmann tail of the Fermi–Dirac distribution.) Therefore we can write for the probability p that in unit time interval an electron is excited from a trap into the conduction band,

$$p = s \exp(-E_b/k_B T) \tag{6.23}$$

where s is called the event frequency and has dimensions of $(\text{time})^{-1}$.

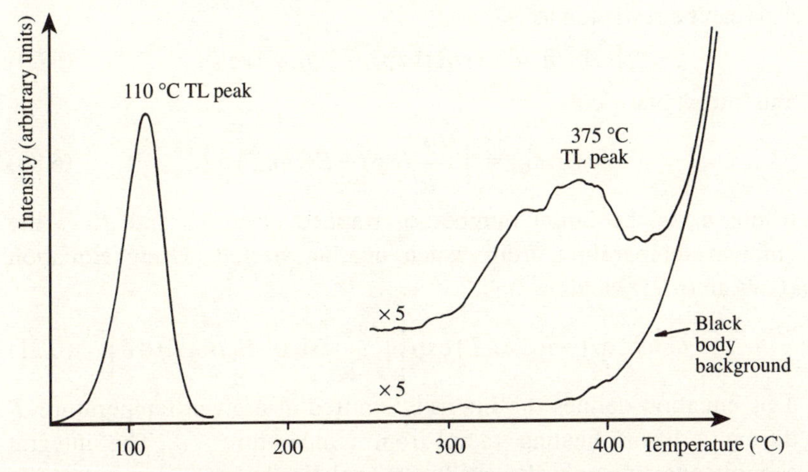

Fig. 6.6 A thermoluminescence 'glow curve' from quartz in Iron Age pottery. Note the peaks at 110 °C and 375 °C corresponding to the emptying of traps of different binding energy. The black body background radiation from the sample after the traps had been emptied is also shown. (Courtesy Mr I. K. Bailiff, Dept of Archaeology, Durham University.)

If the total number of trapped electrons at time t after the heating has started is given by n, then the rate of release of electrons from the traps, $-\mathrm{d}n/\mathrm{d}t$, is given by

$$
\begin{aligned}
-\mathrm{d}n/\mathrm{d}t &= np \\
&= ns\exp(-E_\mathrm{b}/k_\mathrm{B}T). \qquad (6.24)
\end{aligned}
$$

The intensity of light emitted I is proportional to the rate of release of electrons and hence

$$
\begin{aligned}
I &= -C\,\mathrm{d}n/\mathrm{d}t \\
&= Cns\exp(-E_\mathrm{b}/k_\mathrm{B}T) \qquad (6.25)
\end{aligned}
$$

where C is a constant. Let us suppose that the heating rate is such that the temperature T rises linearly with time. Defining R as

$$R = \mathrm{d}T/\mathrm{d}t \qquad (6.26)$$

we have R constant, and this corresponds quite well to the experimental situation.

Therefore, writing

$$\mathrm{d}n/\mathrm{d}t = R(\mathrm{d}n/\mathrm{d}T), \qquad (6.27)$$

Equation (6.24) becomes

$$\mathrm{d}n/\mathrm{d}T = -(ns/R)\exp(-E_\mathrm{b}/k_\mathrm{B}T). \qquad (6.28)$$

This can be rewritten as

$$dn/n = -(s/R)\exp(-E_b/k_BT)\,dT \tag{6.29}$$

and integrating yields

$$\ln(n/n_0) = \int_T^{T_0} \frac{s}{R} \exp(-E_b/k_BT)\,dT \tag{6.30}$$

where n_0 is the initial number of trapped electrons and T_0 is the ambient temperature from which heating started. Using Equation (6.30) in (6.25) yields

$$I = Csn_0 \exp(-E_b/k_BT) \exp\left[\int_T^{T_0} \frac{s}{R} \exp(-E_b/k_BT)\,dT\right]. \tag{6.31}$$

This equation defines the intensity emitted at a given temperature T during a linear heating ramp from temperature T_0. The integral cannot, unfortunately, be evaluated analytically but it can be readily determined numerically. Figure 6.7 shows the result using s = 1 GHz and $E_b = 0.6$ eV for various heating rates. The characteristic low temperature TL peak seen in Fig. 6.6 is rather well modelled by this equation. Note that the position of the peak maximum is a function of heating rate and does not give a direct measure of E_b. One should also note that while the intensity is higher for a faster heating rate, the total number of photons emitted is constant for the three plots. Using

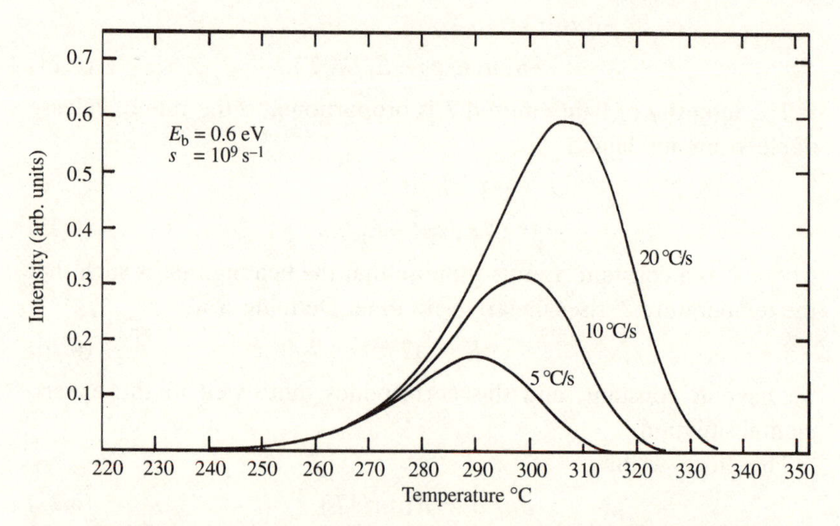

Fig. 6.7 Calculated TL glow curves using Equation (6.31) (known as first order kinetics). Note the movement of the peak maximum as the heating rate is increased.

photon counting techniques, TL glow curves with as few as a total of 600 photons emitted can be recorded before background statistical noise makes measurements unreliable.

6.2.4 Excitons

Before leaving the details of the electron traps and passing on to study the conductivity mechanisms in the important class of extrinsic semiconductors with shallow donor states, we will briefly discuss the concept of an exciton. The quasi-Bohr model can again be used to explain this phenomenon.

When an electron–hole pair is created, for example by absorption of light in a semiconductor, the electron and hole are initially close together in space. They will therefore experience a mutual Coulomb electrostatic force because the electron is negatively charged and the hole is positively charged. This system looks exactly like a miniature hydrogen atom and the quantized energy levels can be determined by use of the Bohr model as with the donor/acceptor states. The major difference is that the electron and hole have comparable masses and we must consider orbits about the centre of mass, rather than about an infinitely heavy nucleus.

For an electron and a hole separated by distance r, the force between them is $e^2/4\pi\varepsilon\varepsilon_0 r^2$ where, as before, ε corresponds to the dielectric constant of the bulk semiconductor. We assume effective masses m_e^* and m_h^* for the electron and hole respectively and distances r_e and r_h from the electron and hole to the centre of mass respectively. Then

$$m_e^* r_e = m_h^* r_h \tag{6.32}$$

and

$$r = r_e[1 + (m_e^*/m_h^*)]. \tag{6.33}$$

From the Bohr model we have

$$m_e^* r_e^2 \omega = n\hbar \tag{6.34}$$

and

$$m_e^* r_e \omega^2 = e^2/(4\pi\varepsilon\varepsilon_0 r^2). \tag{6.35}$$

These equations can be rewritten using Equation (6.33) as

$$Mr^2 \omega = n\hbar \tag{6.36}$$

$$Mr\omega^2 = e^2/(4\pi\varepsilon\varepsilon_0 r^2) \tag{6.37}$$

where M is known as the reduced mass and is given by

$$\frac{1}{M} = \frac{1}{m_e^*} + \frac{1}{m_h^*}. \qquad (6.38)$$

Equations (6.36) and (6.37) are now equivalent to (6.14) and (6.15) so the total binding energy of the electron–hole system is therefore

$$E_n = -e^4 M/2(4\pi\varepsilon\varepsilon_0)^2 n^2 \hbar^2 \qquad (6.39)$$

where n is an integer. This energy is a reduction from the situation where the electron and hole are spatially separated, i.e. with energies corresponding to the bottom of the conduction band and top of the valence band respectively. The interaction therefore gives rise to a set of energy levels just below the bottom of the conduction band into which the electron can be excited by the light which creates the electron–hole pair. That is, we do not need the light photon to have enough energy to create a separated electron–hole pair. Absorption can take place at lower frequency if an excitation is created. The presence of these exciton levels can be seen as structure on the absorption curve when the photon energy is just less than the energy gap. An example of this is shown in Fig. 6.8. We note finally that the

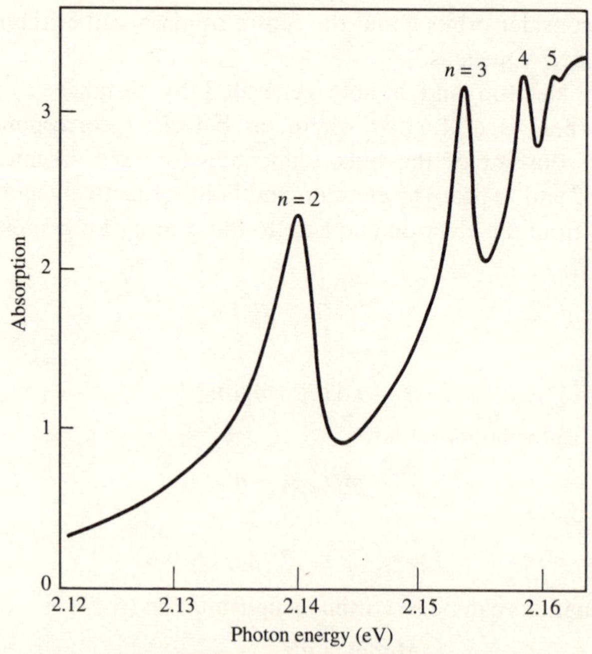

Fig. 6.8 Exciton lines seen in the absorption spectrum of Cu_2O where they are hydrogen-like in their spacing. (After Baumeister, *Phys. Rev.* **121** (1961) 359.)

levels are not discrete, as the electrons and holes are not localized and that, for $m_e^* = m_h^*$, the lowest level lies a distance below the conduction band equal to half the binding energy of the shallow donor states discussed earlier.

6.3 Photoconductivity

A very important application of some semiconductors, for example, CdS, arises from the fact that the electrical resistivity varies with illumination when light is shone onto the material. This variation of electrical resistivity suggests that the number of current carriers is changed by the illumination process and indeed this is what happens. When light falls on a semiconductor, provided that the photon energy is sufficiently high, greater than the band gap E_g, it can be absorbed creating free electrons and free holes. Under a voltage applied across the specimen, these will drift in opposite directions, giving rise to a net current flow. For a constant potential drop across the sample, the current clearly increases with illumination and hence the resistivity falls. We can determine how this current, called the photocurrent varies with illumination using a model similar to that discussed in Section 6.2.3.

Referring always to unit volume of material, suppose that in unit time L photons are absorbed. Let the number of free electrons be n and the number of free holes be p. There is a certain probability that the electrons and holes will recombine, giving out visible light or heat and if the electrons and holes are free, this probability will be proportional to the product of the electron and hole concentration i.e. np. As a result there will be a loss of carriers. Further, some of the free electrons will be trapped in impurity states and the rate of trapping will be proportional to the number of free electrons present. Therefore, we can write an equation for the rate of change in free electron concentration with time, namely

$$\frac{dn}{dt} = L - Anp - Bn \tag{6.40}$$

where A is a recombination-rate proportionality constant and B is a trapping-rate proportionality constant.

Each absorbed photon creates an electron–hole pair and hence $n = p$ and therefore in the steady state

$$L - An^2 - Bn = 0. \tag{6.41}$$

This quadratic equation has a positive solution

$$n = \frac{(B^2 + 4AL)^{1/2} - B}{2A}. \tag{6.42}$$

For small L, the equation can be expanded using the binomial theorem to give

$$n = L/B. \tag{6.43}$$

We can neglect the hole contribution to the photocurrent because holes usually have very much lower mobility than electrons and we see immediately that the photocurrent (proportional to the electron concentration n) varies linearly with illumination at low illumination levels. Here the trapping mechanism dominates.

When L is large we can neglect B and Equation (6.42) becomes

$$n = (L/A)^{1/2} \tag{6.44}$$

that is, the photocurrent varies as the square root of illumination. Fig. 6.9 shows the change in slope with increasing illumination for an ORP12 CdS photoresistor. At high illumination there is particularly good agreement with the prediction of Equation (6.44).

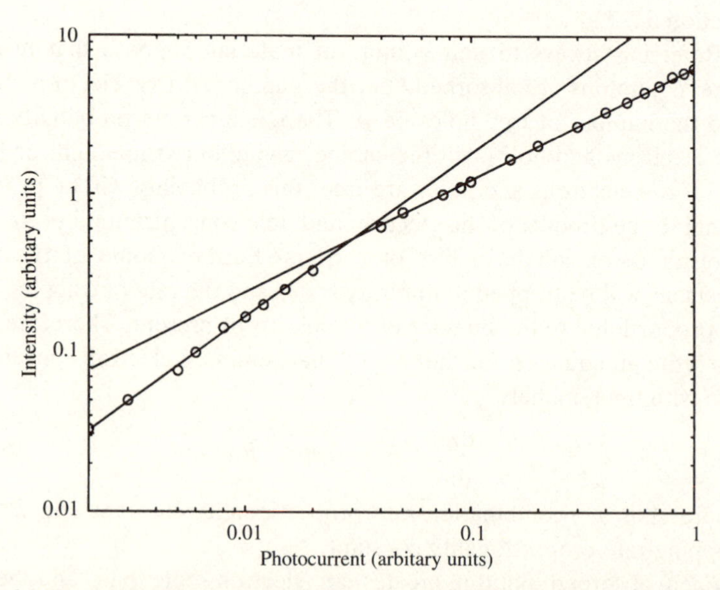

Fig. 6.9 Experimental variation of photocurrent with intensity of illumination for a CdS light-dependent resistor under constant bias. Note the change in slope from 0.71 at low illumination to 0.47 at high illumination.

6.4 Conductivity of extrinsic semiconductors

The presence of shallow donor or acceptor impurities can have a dramatic effect on the electrical properties of a semiconductor. We will use the narrow band model to obtain a quantitative result for the variation of electrical conductivity with dopant concentration and temperature. Suppose that a semiconductor such as silicon has N_d pentavalent (donor) atoms per unit volume. These form localized donor states lying E_b below the conduction band in which there are Z states per unit volume. Let the average energy of the (narrow) conduction band be E_c and the donor ground state energy be E_d (Fig. 6.10).

The number of electrons in the conduction band n_c is given by

$$n_c = \frac{Z}{\exp[(E_c - E_F)/k_B T] + 1} \approx Z \exp[(E_F - E_c)/k_B T] \tag{6.45}$$

if E_F lies greater than about $2k_B T$ from the conduction band. We will examine this approximation later. Similarly the number of electrons n_d in donor states at energy E_d is

$$n_d = \frac{N_d}{\exp[(E_d - E_F)/k_B T] + 1}. \tag{6.46}$$

The number of empty donor states (i.e. ionized donor states) is

$$m_d = N_d - N_d \{\exp[(E_d - E_F)/k_B T] + 1\}^{-1}. \tag{6.47}$$

If the conduction band lies much greater than $k_B T$ from the valence band, the probability of electrons being excited directly from the valence band to the conduction band is negligible. This is a very good approximation at room temperature for silicon where $E_g = 1.1$ eV. Also we can see by consideration of the situation at $T = 0$ where all the donor states are neutral, that the Fermi level lies between the donor levels and the conduction band. Assuming that E_F lies more than about $2k_B T$ from E_d, we can expand Equation (6.47) by the binomial theorem yielding,

$$m_d = N_d \exp[(E_d - E_F)/k_B T]. \tag{6.48}$$

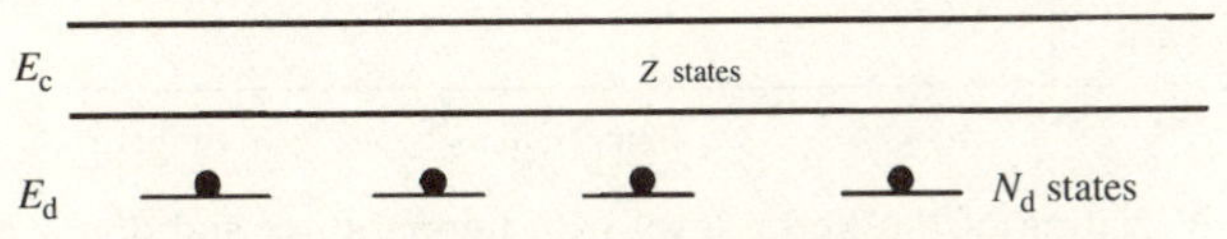

Fig. 6.10 Donor states in the narrow band model.

If the probability of direct excitation across the gap (the intrinsic conductivity) is neglected, the number of electrons in the conduction band n_c must equal the number of ionized donors, m_d. Hence

$$N_d \exp[(E_d - E_F)/k_B T] = Z \exp[(E_F - E_c)/k_B T]. \quad (6.49)$$

This yields

$$E_F = \tfrac{1}{2}(E_c + E_d) + \tfrac{1}{2}k_B T \ln(N_d/Z). \quad (6.50)$$

Figure 6.11 shows how the Fermi level varies with temperature and donor concentration. According to the narrow band model, E_F falls linearly with increasing T. (Note that $\ln(N_d/Z)$ is negative as $N_d \ll Z$ or the impurities will have a greater concentration than the host semiconductor atoms!)

We can use Equation (6.50) in Equation (6.45) to deduce the free carrier concentration. This gives

$$n_c = (ZN_d)^{1/2} \exp(-E_b/2k_B T) \quad (6.51)$$

where $E_b = E_c - E_d$. Two points should be noted. Firstly, the number of carriers increases exponentially with T^{-1} as found for the intrinsic material. However, the gradient is now much less, as $E_b \ll E_g$ and at

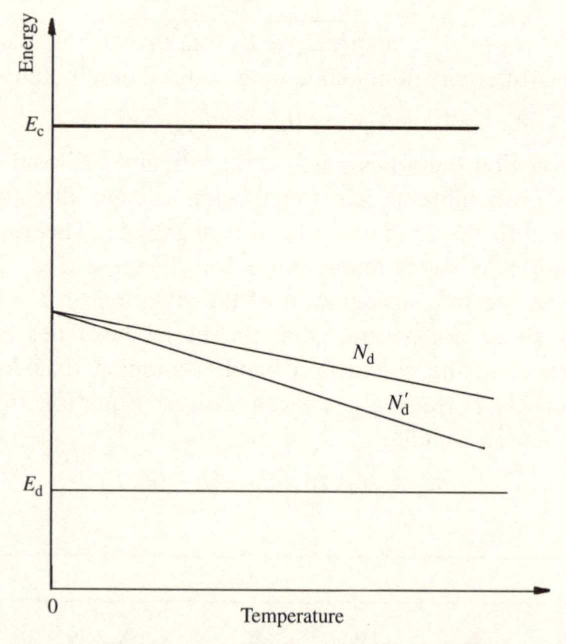

Fig. 6.11 Variation of the Fermi level with temperature and donor concentration. $N_d < N_d'$.

room temperature the exponential term is quite large. Secondly, the carrier concentration at constant temperature depends on the square root of the donor atom concentration. Clearly then, because E_b is so small compared with E_g and comparable at $k_B T$, modest amounts of impurity can give rise to very large changes in carrier concentration, and hence conductivity (see Equation (6.12)). At low temperatures, the temperature variation of the conductivity is dominated by the rise in carriers excited from impurity states. As the temperature rises we reach a plateau region (Fig. 6.12) where intrinsic conductivity is negligible, but almost all of the impurity states are ionized. This is the case for germanium around room temperature. At even higher temperatures, intrinsic conductivity becomes important and the conductivity rises steeply. Plotted as $\ln(n_c)$ versus $1/T$ we find two regions which fit well to straight lines. These have slopes of $-E_b/2k_B$ and $-E_g/2k_B$ respectively in the low and high temperature regions. This provides a method of measuring both E_g and E_b.

While the treatment given above was entirely concerned with donor impurities, an identical argument holds for acceptor impurities. Assuming only acceptor states are present, we find that the number of free holes in the valence band p_v is given by

$$p_v = (ZN_a)^{1/2} \exp(-E_b/2k_B T), \tag{6.52}$$

where N_a is the number of acceptor per unit volume. The expression is almost identical to (6.51). In this case the Fermi level lies mid-way between the acceptor levels and the valence band, rising linearly with increasing temperature.

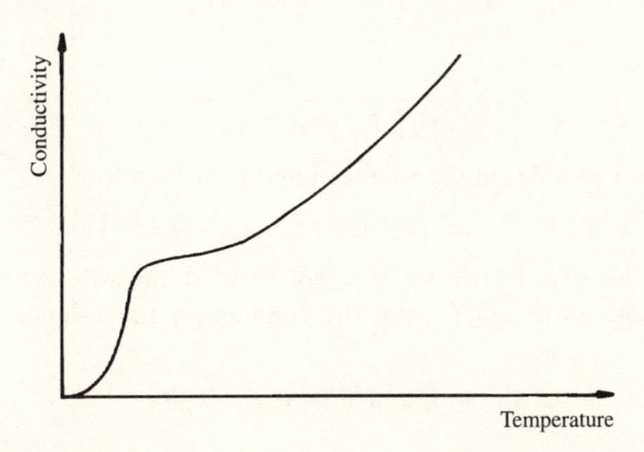

Fig. 6.12 Schematic diagram of the variation in conductivity with temperature of an extrinsic semiconductor.

It is worth noting that in the plateau region, the conductivity of the semiconductor actually falls, because the mobility, μ is determined by the scattering from the (vibrating) crystal lattice. This scattering rises with temperature, and the mobility varies roughly as $T^{-3/2}$, giving rise to a fall in conductivity with temperature in the plateau region.

Because the carrier concentration in the extrinsic region varies so very rapidly with temperature, semiconductors can be used as thermometers. Very precise low temperature thermometry is carried out by measuring the resistance of germanium, and devices to measure temperature at around room temperature are marketed under the name of thermistors.

6.4.1 Compensation

In the previous section we considered semiconductors with only one type of impurity atom present. If both donor and acceptor atoms are present the situation becomes more complex. Surprisingly perhaps, increasing impurity concentration does not always lead to an increase in free carrier concentration. In order to see the origin of this effect, let us consider a semiconductor with an arbitrary concentration of donor and acceptor atoms. Let there be equal numbers of states Z in both conduction and valence bands. Then, quite generally we can write for n_c the number of electrons in the conduction band and n_v the number of electrons in the valence band,

$$n_c = \frac{Z}{\exp[(E_c - E_F)/k_B T] + 1} \tag{6.53}$$

and

$$n_v = \frac{Z}{\exp[(E_v - E_F)/k_B T] + 1}. \tag{6.54}$$

The number of holes in the valence band p_v is therefore

$$p_v = Z - Z\{\exp[(E_v - E_F)/k_B T] + 1\}^{-1}. \tag{6.55}$$

Provided that E_F, which we have not defined and will not do so, lies greater than about $2k_B T$ from the conduction and valence bands we can write

$$n_c \approx Z \exp[(E_F - E_c)/k_B T] \tag{6.56}$$

and

$$p_v \approx Z \exp[(E_v - E_F)/k_B T]. \tag{6.57}$$

Therefore

$$n_c p_v \approx Z^2 \exp[(E_v - E_c)/k_B T] \\ \approx Z^2 \exp(-E_g/k_B T). \quad (6.58)$$

This is a most important result. It shows that the product of free electrons and holes is independent of the impurity concentration. $n_c p_v$ is the same for an intrinsic semiconductor as for one with a heavy concentration of impurities. In the intrinsic semiconductor there are equal numbers of free electrons and holes. Let this number be n_i at a given temperature. Thus

$$n_c p_v = n_i^2 \quad (6.59)$$

at any given temperature. From this result we see that any increase in the free electron concentration must result in a decrease in the free hole concentration and vice versa. This is known as compensation.

Compensation can be used to *reduce* the conductivity of a material by *addition* of impurity. Take for example a hypothetical material having a free electron concentration $n_c = 1000$ and free hole concentration $p_v = 10$. The total free carrier concentration is $n_c + p_v = 1010$. Suppose we add acceptor atoms to increase the free hole concentration a small amount to $p'_v = 20$. The product $n'_c p'_v$ must remain still equal to $n_c p_v$, namely 10 000. Therefore $n'_c = 500$ and the total carrier concentration $n'_c + p'_v = 520$. In other words, addition of a small amount of the minority impurity can drastically reduce the number of free carriers and hence the conductivity. This is most important in integrated circuit device fabrication, as one can build a high resistance region into a low resistance material by *addition* of impurity. It would be impossible to refine the material to get impurities out of a specific region.

6.5 Wide band model

The above results have been developed on the simplest model we can find of a semiconductor, namely that the width of the energy bands is small compared to the energy gap. While it turns out that this is a good approximation we must justify the use of the model. From a study of the $E-k$ curve relating to the nearly free electron approximation (Fig. 4.8(*b*)) we note that at the Brillouin zone boundary (i.e. the energy gap) $\partial E/\partial k = 0$. If we expand the energy in a Taylor series away from that turning point we find that the curve is, to a first approximation, parabolic with respect to the band edge. This leads directly to a density of states near the band edges which varies as

$A(E - E_c)^{1/2}$ in the conduction band and $B(E_v - E)^{1/2}$ in the valence band, where E_c and E_v are the energies of the respective band edge. Comparison with Equation (2.34) gives

$$A = (2m_e^{*3})^{1/2} V/\pi^2\hbar^3 \tag{6.60}$$

and

$$B = (2m_h^{*3})^{1/2} V/\pi^2\hbar^3 \tag{6.61}$$

where m_e^* and m_h^* are the electron and hole effective masses and V is the sample volume.

We can proceed as before to calculate the carrier concentration in an intrinsic semiconductor. The number of electrons in the conduction band is now given by the Fermi–Dirac distribution multiplied by the density of states integrated over the whole band. This is

$$n_c = \int_{E_c}^{\infty} \frac{A(E - E_c)^{1/2}\, dE}{\exp[(E - E_F)/k_B T] + 1}. \tag{6.62}$$

We have assumed that the upper limit of the integral is infinity and indeed this is a satisfactory approximation as there is a negative exponential term in E. Provided that the energy lies greater than about $2k_B T$ from the conduction band edge we have

$$n_c = \int_{E_c}^{\infty} \frac{A(E - E_c)^{1/2}\, dE}{\exp(E - E_F)/k_B T}, \tag{6.63}$$

and writing $x = (E - E_c)/k_B T$ yields

$$n_c = A(k_B T)^{3/2} \exp[(E_F - E_c)/k_B T] \int_0^{\infty} x^{1/2} e^{-x}\, dx. \tag{6.64}$$

The integral is totally independent of T and is a pure number. In fact it is a standard integral and equal to $\pi^{1/2}/2$. Thus

$$n_c = \tfrac{1}{2} A \pi^{1/2} (k_B T)^{3/2} \exp(E_F - E_c)/k_B T. \tag{6.65}$$

Similarly the number of electrons in the valence band n_v is given by

$$n_v = \int_{-\infty}^{E_v} \frac{B(E_v - E)^{1/2}\, dE}{\exp[(E - E_F)/k_B T] + 1}. \tag{6.66}$$

Provided that the Fermi level is greater than $2k_B T$ from the valence band edge we can write for the number of holes in the valence band p_v,

$$\begin{aligned} p_v &= \int_{-\infty}^{E_v} B(E_v - E)^{1/2} (1 - \{\exp[(E - E_F)/k_B T] + 1\}^{-1})\, dE \\ &\approx \int_{-\infty}^{E_v} B(E_v - E)^{1/2} \exp[(E - E_F)/k_B T]\, dE. \end{aligned} \tag{6.67}$$

Writing $y = (E_v - E)/k_B T$ yields

$$p_v = B(k_B T)^{3/2} \exp[(E_v - E_F)/k_B T] \int_0^\infty y^{1/2} e^{-y} \, dy$$
$$= \tfrac{1}{2} B \pi^{1/2} (k_B T)^{3/2} \exp(E_v - E_F)/k_B T. \quad (6.68)$$

As $n_c = p_v$ we have

$$E_F = \left(\frac{E_c + E_v}{2}\right) + aT \quad (6.69)$$

where $a = \frac{1}{2} k_B \ln(B/A)$, independent of temperature. Equation (6.69) should be compared with Equation (6.7) based on the narrow band model. We see that the wide band model gives a Fermi energy which varies linearly with temperature. Substitution of Equation (6.69) in (6.65) gives

$$n_c = \tfrac{1}{2}(AB\pi)^{1/2} (k_B T)^{3/2} \exp(-E_g/2k_B T). \quad (6.70)$$

This result is very similar to that derived from the narrow band model except for the additional temperature dependent factor in $T^{3/2}$. The exponential varies much faster than the power factor, however, and a $\ln(n_c)$ versus $1/T$ plot appears linear.

The main reason why the narrow band model works so well is that the majority of electrons in the conduction band lie in the narrow energy range of about $k_B T$ from the band edge. As the energy rises, so the probability of occupation decreases exponentially and the contribution to the carrier concentration becomes negligible. We have in effect, two narrow bands of carriers close to the band edges. The difference between this and the wide band model is essentially that in one the effective density of states in the band is a function of temperature rather than a constant.

6.6 Majority and minority carriers

In the discussion so far we have mainly considered the problem from a viewpoint of a semiconductor containing only one type of current carrier. For example, in the semiconductor with only donor atoms present, we considered only the extrinsic conduction associated with these donor electrons. However, many materials have both acceptors and donors present and in any case, the result of Equation (6.59) must hold irrespective of impurity concentration. In other words, in a material containing predominantly donors, which is known as n-type, although the free electron concentration will be high there will always be a small, but non-negligible concentration of holes. Similarly in a material with a predominance of acceptor atoms, known as p-type, there will be a majority of holes but a small number of free conduction electrons. These minority carriers, that is holes in n-type material and electrons in p-type material, are of utmost importance in the operation of devices.

Despite there being a well defined minority carrier density, we must remember that it is only an average number and corresponds to a situation of dynamic equilibrium. Electrons and holes have quite a high probability of being attracted together by the Coulomb interaction and recombining. This recombination process can, as we have seen, lead to light emission and indeed form the basis of the operation of the light emitting diode. In order to keep the average minority carrier concentration constant, there will be new carriers created by thermal excitation.

The average distance a minority carrier can travel through a material before recombination takes place is very important to the operation of a number of devices including the p–n–p and n–p–n junction transistors. It is related to both the lifetime τ_r, that is the average time before recombination, and the minority carrier mobility μ. Minority carrier mobility can be measured by specifically injecting a pulse of minority carriers into a rod of semiconductor and measuring the time taken for the pulse of charge to be swept down the specimen in various applied bias fields. The experiment is shown schematically in Fig. 6.13. Electron–hole pairs are created at one end by a pulsed light source focussed onto the rod. The pulsing signal is used to trigger an oscilloscope whose *y* plates are connected to the detector, which is a reverse

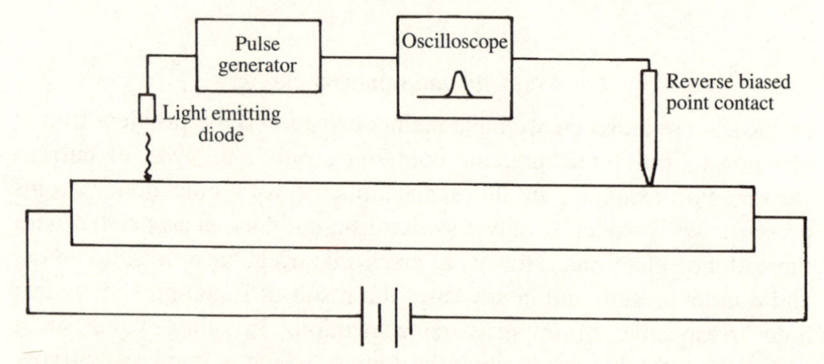

Fig. 6.13 Schematic diagram of experimental arrangement to measure minority carrier mobilities and lifetimes.

biased point metal contact on the other end of the specimen. Under a bias voltage applied across the ends of the specimen rod, the minority carriers drift towards the detector and their arrival is detected as a now diffuse peak on the oscilloscope trace. The time to travel down the rod is measured directly from the oscilloscope trace. It is found that the time taken to travel to the detector is inversely proportional to bias voltage, hence demonstrating the existence of a well defined carrier mobility. If $\langle v \rangle$ is the drift velocity of the minority carriers in electric field $\mathscr{E}$, then the time t taken to travel distance x is

$$t = x/\langle v \rangle = x/\mu\mathscr{E}. \tag{6.71}$$

We can estimate the recombination time by examining the area under the oscilloscope trace pulse as the detector is moved along the rod. An exponential decay in the electrical pulse is found, indicating a constant probability of recombination per unit time interval. The fraction of minority carriers ($\mathrm{d}q/q$) recombining with majority carriers in time $\mathrm{d}t$ is then

$$\mathrm{d}q/q = -\mathrm{d}t/\tau_{\mathrm{r}}. \tag{6.72}$$

This equation yields

$$\begin{aligned} q &= q_0 \exp(-t/\tau_{\mathrm{r}}) \\ &= n_0 e \exp(-t/\tau_{\mathrm{r}}) \end{aligned} \tag{6.73}$$

where n_0 is the number of minority carriers initially created. From Equation (6.71) we then have

$$q = n_0 e \exp(-x/\mathscr{E}\mu\tau_{\mathrm{r}}) \tag{6.74}$$

and so by measuring the decay of the charge pulse together with the mobility we can determine the recombination time τ_{r}.

Problems

6.1 An intrinsic semiconductor has a band gap of 3 eV and Z levels in conduction and valence bands. A dopant is introduced giving N_{d} levels 0.1 eV below the conduction band. Calculate the value of N_{d}/Z needed to make the conductivity 100 times the intrinsic value at room temperature, assuming all carrier mobilities are the same.

6.2 The Hall coefficient of a certain specimen of silicon was found to be $-7.35 \times 10^{-5}\,\mathrm{m^3\,C^{-1}}$ from 100 to 400 K. State whether this specimen is intrinsic or extrinsic at room temperature and, if extrinsic, state whether n-type or p-type. The electrical conductivity at room temperature is found to be 200 $(\text{ohm m})^{-1}$. Calculate the density and mobility of the charge carriers.

6.3 What is the difference in the Fermi level energy (in eV) at 300 K of two specimens of a semiconductor of element X, doped with an element Y which introduces donors, if the concentration of Y in the two specimens is in the ratio of 1:3?

6.4 Close to the conduction band edge, E_c, assume that the conduction band carrier density $N(E)$ is proportional to $(E - E_c)^{1/2} \exp[-(E - E_F)/k_B T]$. Show that $N(E)$ has a maximum at $k_B T/2$ above E_c.

6.5 The effective mass of an electron in indium antimonide is very low, namely 0.014 of the free electron mass. The dielectric constant is 17 and the energy between valence and conduction bands is 0.23 eV. If it is suitably doped to provide donor levels, calculate the donor ionization energy in eV. Note that this is very small compared with $k_B T$ at room temperature. Sketch the expected variation of conductivity with temperature. What would be the slope of a plot of log (conductivity) versus $1/T$ at (a) 4.2 K and (b) 600 K?

Calculate the radius of the ground state Bohr orbit of the donor electron. Estimate at what donor concentration the orbits of adjacent donor atoms appreciably overlap. (This tends to produce an impurity BAND – as electrons can move from one donor atom to another, presumably by a hopping mechanism. Such effects are not important in semiconductors with larger effective electron masses.)

6.6 In an intrinsic semiconductor, the top of the valence band defines $E = 0$ and the bottom of the conduction band is at E_g. The density of electron states between E and $E + dE$ in the conduction band is $4\pi(2m_e^*/\hbar^2)^{3/2}(E - E_g)^{1/2}\,dE$. Similarly, the density of hole states is $4\pi(2m_h^*/\hbar^2)^{3/2}(-E)^{1/2}\,dE$. If $E_g = 1$ eV and the Fermi energy lies 0.48 eV above the top of the valence band, determine the ratio of the electron and hole masses, m_e^*/m_h^*, at a temperature of 300 K.

6.7 A semiconductor has a narrow conduction band containing Z states. It has donor states, only a small fraction f of which are filled at $T = 0$. Show that the density of free electrons in the conduction band n_c at temperature T is

$$n_c = Zf \exp(-\Delta E/k_B T)$$

where ΔE is the donor ionization energy. Neglect excitation from the valence band, and assume $\Delta E \gtrsim 2k_BT$.

6.8 A certain insulator has Z states in the conduction band, a large number N of electron traps at energy ΔE below the bottom of the conduction band and also N donor states at energy $2\Delta E$ below the bottom of the conduction band. At temperature $T = 0$, all donor states are filled and all the traps are empty.

Determine the position of the Fermi level assuming the narrow band model. Prove that the number of electrons n_c in the conduction band at temperature T is

$$n_c = Z \exp(-3\Delta E/2k_BT).$$

6.9 A specimen of ancient pottery is heated at a linear rate from 300 K. In the low temperature tail of the glow curve peak, the intensity varied with temperature as shown below:

Temperature (K)	350	360	370	380	390
Intensity (cps)	640	1010	1560	2350	3480

Determine the binding energy of the electron traps responsible for the TL.

6.10 A semiconductor is satisfactorily represented by the narrow band model, with Z states in the valence band. The conduction band may be considered to be a large energy away from the valence band, which is taken to be at $E = 0$. Just above the valence band lie N_1 acceptor levels at energy E_1 and N_2 levels at energy E_2. Derive expressions for the concentration of holes in the valence band in the two cases where

(a) $N_1 \ll N_2$ and $E_2 > E_1$.
(b) $N_1 \approx N_2$ and $E_2 \gg E_1$.

Assume that the acceptor states lies more than $2k_BT$ away from the valence band.

Explain qualitatively the reason for these two results.

6.11 The energy gap of intrinsic GaAs is 1.5 eV. Determine the ratio of the electron concentrations at 300 K and 600 K, assuming the wide band model and that m_e^* and m_h^* are temperature independent.

7
Semiconductor devices

One of the most remarkable developments of the last decade has been the growth of the semiconductor industry. The exploitation of the properties of semiconducting materials to build large logic arrays on a single piece of crystal has led to a dramatic increase in the processing power and memory capacity of small computers, with a simultaneous fall in price. Development of single chip microprocessors has revolutionized our mode of working in a whole range of fields from the factory floor to the office. In this chapter we will examine the basic physics associated with a number of devices, and as many devices rely on the properties of junctions between n- and p-type semiconductors a fairly detailed discussion of such junctions is given before individual devices are treated. However, because all devices require connection to metallic wires in order to join components together we will first examine the metal–semiconductor junction. As it turns out this proves to be an excellent introduction to the p–n junction as well as providing a glimpse into some of the not-so-obvious pitfalls associated with device manufacture.

7.1 Metal–semiconductor junctions

7.1.1 The Schottky barrier

Let us suppose that a piece of metal is brought into contact with a piece of n-type semiconductor. (Of course, in practice, the metal would be evaporated on the semiconductor as a thin film, or attached with a soldered connection in which an alloy is formed, but such a naïve picture is useful to fix our ideas.) In general the Fermi levels of the two materials will be at different energies (Fig. 7.1(a)). As with the metal–metal junction, this is not a minimum energy situation and

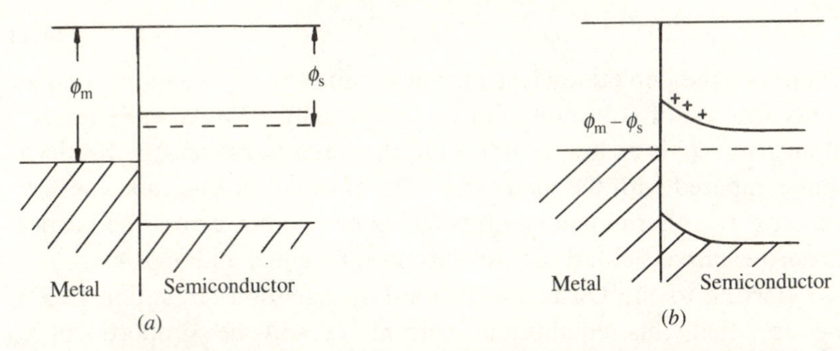

Fig. 7.1 Metal–semiconductor junction. (*a*) Fermi levels and work function for separate metal and semiconductor. (*b*) Formation of depletion region and resulting band bending due to electron transfer from semiconductor to metal.

electrons flow. Let ϕ_s be the effective work function of the semiconductor and ϕ_m the work function of the metal. Then if $\phi_m > \phi_s$ electrons will flow from the semiconductor into the metal. The electron density in the semiconductor is very much lower than in the metal and consequently in order to equate the Fermi levels, a substantial volume of semiconductor becomes depleted of charge carriers. This region is known as the *depletion layer* and we meet it again in the discussion of the p–n junction. Due to the relatively low carrier concentration this depletion layer is quite thick, in contrast to the metal–metal interface and this implies that electrons have a very low probability of quantum mechanical tunnelling through the barrier. As in the p–n junction, carriers cross the barrier by diffusion.

The energy barrier (Fig. 7.1(*b*)) can be seen to be of height $\phi_m - \phi_s$ and arises from the presence of an electrostatic dipole layer between the electrons transferred to the metal and the ionized donor states in the depletion layer of the semiconductor. Clearly charge transfer occurs until the electrostatic potential associated with the dipole layer is just sufficient to bring the Fermi levels in the metal and semiconductor into coincidence. We can determine the barrier thickness known as the Schottky barrier, by solving Poisson's equation with the appropriate boundary conditions.

The potential energy ϕ of an electron at position x in the depletion region is given by Poisson's equation, namely,

$$d^2\phi/dx^2 = n_d e^2/\varepsilon\varepsilon_0 \quad (7.1)$$

where n_d is the donor concentration, and we assume that all donors in the region $x = 0$ to $x = x_0$ are ionized. If $\phi = 0$ at $x = 0$ and $\phi = \phi_m - \phi_s$ at $x = x_0$ we have on integration,

$$\phi_m - \phi_s = n_d x_0^2 e^2/\varepsilon\varepsilon_0. \qquad (7.2)$$

Therefore the depletion layer thickness varies as $n_d^{-1/2}$ and for a donor concentration of 10^{17} atoms/cm^3, x_0 is typically 1000 Å, while for 10^{19} atoms/cm^3, x_0 is as low as 100 Å. In the latter situation, the barrier is thin compared with the mean free path of the electrons and, in effect, we have two thermionic emitters facing each other across the barrier. In zero external field, these currents must be equal and opposite.

Referring to the Dushman–Richardson equation (Equation (3.32)) we see that this equilibrium current I_0 will be proportional to $\exp[(\phi_s - \phi_m)/k_B T]$. Let us write

$$I_0 = A \exp[(\phi_s - \phi_m)/k_B T]. \qquad (7.3)$$

On application of a forward bias voltage V so that the semiconductor becomes (conventionally) negative with respect to the metal, the Fermi level of the semiconductor is raised with respect to the Fermi level of the metal. The current from semiconductor to metal will now be increased as the effective work function is lowered. This current I_{SM} is

$$I_{SM} = A \exp[(\phi_s - \phi_m + eV)/k_B T]$$
$$= I_0 \exp(eV/k_B T). \qquad (7.4)$$

The current from metal to semiconductor remains constant as the work function remains $\phi_m - \phi_s$ and this current I_{MS} is equal to I_0. Thus the net current flowing I, is given by

$$I = I_0[\exp(eV/k_B T) - 1]. \qquad (7.5)$$

It is easy to show, similarly, that in reverse bias, $-V$, the current is

$$I = I_0[1 - \exp(-eV/k_B T)]. \qquad (7.6)$$

Figure 7.2 shows this current–voltage (I–V) characteristic. Clearly the reverse current is very small in comparison with the forward current and the junction acts as a rectifier. This rectifying action can be exploited to produce a diode and point contact diodes are used extensively where low capacitance is required, for example in the detection of 3 cm microwaves. However, it is a serious nuisance when one is trying to make contacts between metal wires and semiconductor components. If the model discussed above were the whole story, in order to make an ohmic (non-rectifying) contact between a metal and an n-type semiconductor, we would need to select a metal with a work function less than the effective work function of the semiconductor. Then, in order to equate Fermi levels, there would have to be excess

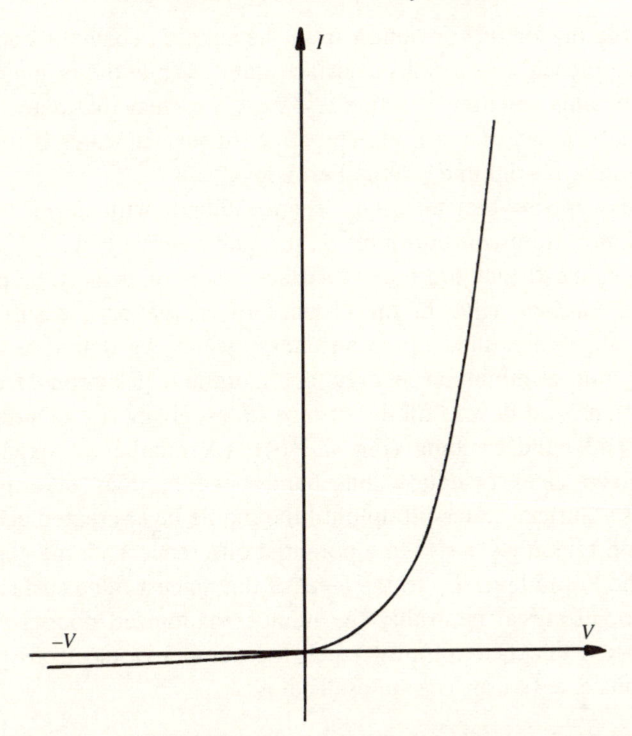

Fig. 7.2 $I-V$ characteristic of an ideal metal–semiconductor junction.

negative charge in the junction region. The electrons transferred from the metal would be mobile, unlike the static positive ion cores of Fig. 7.1 and application of an electric field of either sense would give an instability resulting in current flow. Similarly ohmic contacts between metals and p-type semiconductors require the metal work function larger than the effective work function of the semiconductor.

7.1.2 Surface states

This simple model of the Schottky barrier diode does not, however, give a complete description of the behaviour of metal–semiconductor junctions. Indeed, until a few years ago, this the simplest of semiconductor devices was probably the least well understood. The problem is that not all metal–semiconductor junctions expected to show rectifying properties do so. Further, Al–GaAs and Au–GaAs junctions give the same value for the height of the barrier despite a 20% difference in the work functions of the two metals.

The reason for this deviation from the simple behaviour appears to lie in the presence of localized surface states. While the origin of these states remains a matter of active research, it is clear that defects in the crystal lattice can play a role. The effect of surface states is to reduce the Schottky barrier and pin the Fermi level.

Let us suppose that an n-type semiconductor with Fermi level E_F^0 has a donor atom concentration n_d per unit volume (Fig. 7.3(*a*)). Let us also suppose that there exist surface states of density n_s per unit area per *electron volt*. In the absence of charge on the surface the surface states are filled up to an energy which we define as $E_0 = 0$. This is not a minimum energy configuration. Electrons from the conduction band flow to fill the surface states, giving rise to a depletion region and band bending (Fig. 7.3(*b*)). (A number of experiments have given clear evidence that band bending does take place at surfaces.) Surface states fill up until the dipole layer created across the depletion region gives rise to a potential difference V_0 large enough to bring the Fermi level E_F to the level of the highest filled surface state. Then for electrical neutrality the number of ionized donors per unit area over a thickness x_0 must equal the number of electrons per unit area transferred to surface states. That is

$$n_s E_F = n_s(E_F^0 - eV_0) = n_d x_0. \tag{7.7}$$

Now from Equation (7.2) we have

$$V_0 = n_d x_0^2 e/2\varepsilon\varepsilon_0. \tag{7.8}$$

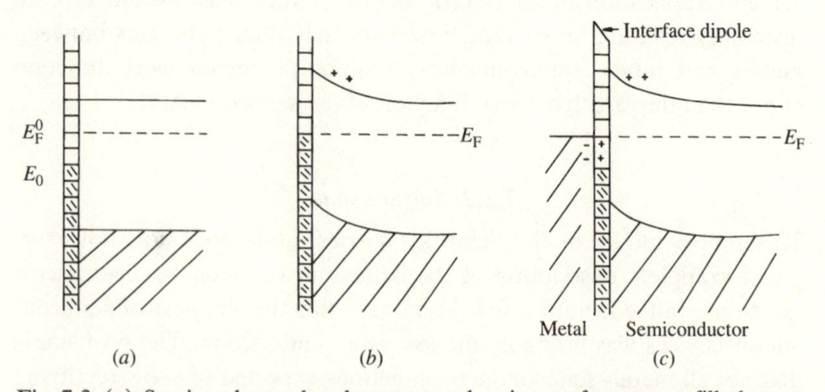

Fig. 7.3 (*a*) Semiconductor band structure showing surface states filled up to energy E_0. (*b*) Band bending close to the surface to equalize Fermi levels. (*c*) Band structure of a metal–semiconductor junction with surface states, showing interface dipole.

Hence,

$$n_s^2 e(E_F^0 - eV_0)^2 = 2\varepsilon\varepsilon_0 n_d V_0 \quad (7.9)$$

If n_s is small, the only way to make this equation balance for modestly doped materials is for V_0 to be small.

Now let us consider the effect of evaporating a metal onto the surface. Electrons are transferred from the surface states into the metal until the Fermi level of the metal is equal to the topmost filled surface state. Because the surface states screen the electrons in the semiconductor, there is no bulk transfer of charge from the semiconductor (Fig. 7.3(*c*)). As the surface states extend over only a very narrow depth the dipole layer produced by the transfer of charge to the metal is very thin. Electrons are able to tunnel through this barrier and no rectifying action takes place. (It is for a similar reason, namely that the dipole layer extends only over a very short distance, that metal–metal junctions are non-rectifying. Again electrons can readily tunnel through the barrier.) As we have seen from Equation (7.9), the height of the barrier eV_0 due to the depletion layer, thickness x_0, can be very small and thus by inclusion of the surface states, we see how metal–semiconductor junctions can be ohmic. A very dramatic example of how a very thin coating of aluminium between gold and the semiconductor CdSe can alter the junction properties from rectifying to ohmic is illustrated in Fig. 7.4. The presence of the aluminium introduces surface states which have a major effect on the junction properties. An excellent review of this, the most enigmatic of junctions, has been given by R. H. Williams in *Contemporary Physics* **23** (1982) 329–51.

7.2 The p–n semiconductor junction

7.2.1 The depletion layer

The physics of the p–n junction is in many ways similar to that of the metal–semiconductor junction except that we now have two types of carrier to consider, electrons and holes. As we recall from Chapter 6, an n-type semiconductor has a majority of free electrons, a p-type semiconductor has a majority of free holes. Let us suppose that two pieces of semiconductor, one n-type and one p-type are brought into contact. The situation, illustrated in Fig. 7.5(*a*), is unstable. Electrons and holes close to the interface will recombine leaving a layer on either side of the interface depleted of free carriers. As evident in Fig. 7.5(*b*) this depletion layer is an electrostatic dipole layer formed between the (static) positively charged donor ions on the n-side and the negatively

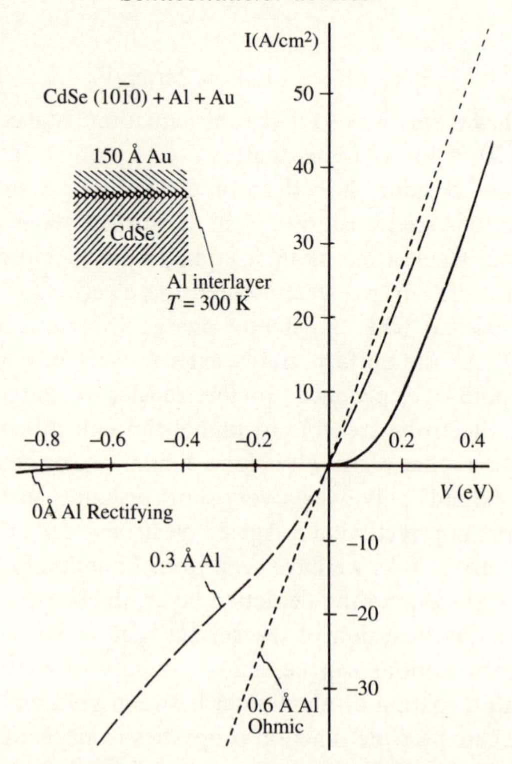

Fig. 7.4 Junction characteristic of a Au–CdSe junction with and without a thin layer of Al at the interface. The Al changes the junction from rectifying to ohmic. (From R. H. Williams, *Contemparory Physics* **23** (1982) 329–51, After C. F. Brucker and L. K. Brillson, *Appl. Phys. Lett.* **39** (1981) 67.)

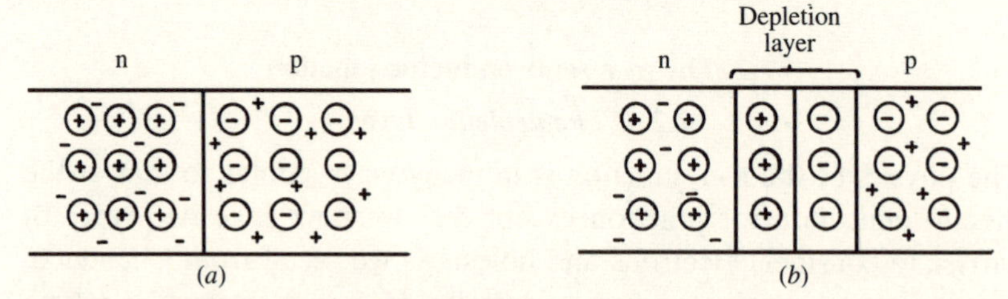

Fig. 7.5 (*a*) Real space representation of p- and n-type material. (*b*) Formation of the depletion layer at the junction.

charged acceptor ions on the p-side. In exactly the same manner as with the metal–semiconductor junction, the electrostatic potential across this dipole layer leads to a equalization of the Fermi levels. We will return to the description after a brief examination of the electrostatics of the depletion layer.

7.2.2 Electrostatics of the depletion layer

Again, the potential at any point can be determined by solution of Poisson's equation

$$d^2V/dx^2 = -\rho/\varepsilon\varepsilon_0 \quad (7.10)$$

where ρ is the density of the static ionic charges. Let this density be ρ_d and ρ_a in the n- and p-sides of the depletion layer respectively, and let $x = 0$ at the junction with the depletion layer extending to $-l_n$ and $+l_p$ on either side. Also let us assume that the junction is abrupt and there is no concentration gradient of donors or acceptors in the depletion region.

Then integration of Poisson's equation gives for the n-side

$$\varepsilon\varepsilon_0 dV/dx = -\rho_d x + \text{constant}. \quad (7.11)$$

At $x = -l_n$, $dV/dx = 0$ and thus

$$\varepsilon\varepsilon_0(dV/dx)_{x=0} = -\rho_d l_n. \quad (7.12)$$

Now similarly integration gives, for the p-side,

$$\varepsilon\varepsilon_0 dV/dx = -\rho_a x + \text{constant} \quad (7.13)$$

and

$$\varepsilon\varepsilon_0(dV/dx)_{x=0} = \rho_a l_p. \quad (7.14)$$

Combining (7.12) and (7.14) yields

$$\rho_a l_p = -\rho_d l_n. \quad (7.15)$$

From this we can see that the effect of very heavy doping on one side of the junction is to give a very narrow depletion layer on the opposite side. This has important implications in the design of tunnel diodes for example.

Integration once again yields

$$\varepsilon\varepsilon_0(V_0 - V_n) = -\tfrac{1}{2}\rho_d l_n^2 \quad (7.16)$$

and

$$\varepsilon\varepsilon_0(V_p - V_0) = \tfrac{1}{2}\rho_a l_p^2 \quad (7.17)$$

where V_n and V_p are the potentials in the n- and p-type regions respectively and V_0 is the potential at the junction (at $x = 0$). The potential difference V across the junction is then

$$V = (\rho_a l_p^2 - \rho_d l_n^2)/2\varepsilon\varepsilon_0 \quad (7.18)$$

or, using (7.15),

$$V = \rho_d l_n^2[(\rho_d/\rho_a) - 1]/2\varepsilon\varepsilon_0. \quad (7.19)$$

If we make the assumption that, in the depletion layer, all donor and acceptor atoms are ionized, we can write

$$V = eN_d l_n^2[(N_d/N_a) + 1]/2\varepsilon\varepsilon_0 \qquad (7.20)$$

where N_d and N_a are the numbers of donor and acceptor atoms per unit volume in the n- and p-type materials. (Note the sign change.)

Equation (7.20) is a very important result in that it is as applicable to the junction under an external bias voltage as it is to zero bias. We made no assumptions in the derivation about the external bias. Thus we see that the width of the depletion layer is a function of bias voltage. As the depletion layer is a dipole layer, varying the width varies the capacitance (for such it is) of the layer. The p–n junction thus acts as a variable capacitor. Now the charge Q per unit area due to the dipole layer is given by

$$Q = -eN_d l_n, \qquad (7.21)$$

and as the capacitance C is given by

$$C = \mathrm{d}Q/\mathrm{d}V \qquad (7.22)$$

we have after a little re-arrangement

$$C = \tfrac{1}{2}V^{-1/2}\left[2\varepsilon\varepsilon_0 e\left(\frac{N_a N_d}{N_a + N_d}\right)\right]^{1/2}. \qquad (7.23)$$

Although the p–n junction must be used in reverse bias so that no significant d.c. current flows, it is extensively used in the form of a variable voltage capacitor in a number of frequency locking and frequency modulation circuits.

7.2.3 The p–n junction rectifier

The effect of the potential difference set up across the depletion layer is to change the relative positions of the Fermi levels on either side of the junction. Before the depletion layer is set up the band structure is schematically shown as in Fig. 7.6(*a*) while after the depletion layer is set up, the Fermi levels are equal (Fig. 7.6(*b*)). Note that the voltage across the junction is almost equal to the energy gap.

The number of electron carriers on the p-type side of the junction is small and we can consider them to occupy levels very close to the bottom of the conduction band. Further, while the electron carrier concentration on the n-type side is very much greater than on the p-type side, we can still consider most of the carriers to occupy energy levels close to the bottom of the conduction band. (This is a good

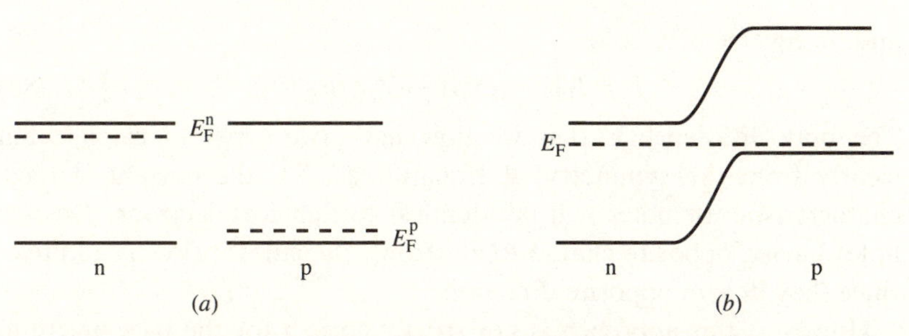

Fig. 7.6 (*a*) Band diagrams of n- and p-type semiconductors. (*b*) Band bending at the interface of a p–n junction as the Fermi levels are equated.

approximation due to the rapid fall in the magnitude of the Fermi–Dirac distribution when $(E - E_F) > 2k_BT$.)

Let n_n and n_p be the respective free electron concentration on the n- and p-sides respectively, and p_p and p_n be the equivalent hole concentrations. Then, quite generally we have

$$n_p/n_n = p_n/p_p = \exp(-eV_j/k_BT) \tag{7.24}$$

where V_j is the potential drop across the junction. Equation (7.24) is valid whether or not an external bias exists. Let V_B be the potential drop in zero bias and n_{p0}, n_{n0}, p_{p0} and p_{n0} be the electron and hole concentrations in zero bias. Then

$$n_{p0}/n_{n0} = p_{n0}/p_{p0} = \exp(-eV_B/k_BT). \tag{7.25}$$

Because n_n and p_p are very large, they do not change very much on application of a bias voltage. Thus Equation (7.24) is approximately given by

$$n_p/n_{n0} = p_n/p_{p0} = \exp(-eV_j/k_BT). \tag{7.26}$$

If we divide Equation (7.26) by (7.25) we have

$$n_p/n_{p0} = p_n/p_0 = \exp(eV_{ext}/k_BT) \tag{7.27}$$

where $V_{ext} = V_B - V_j$, the externally applied bias voltage. The sense of this bias is chosen such that in *forward bias* the electron concentration on the p-side of the junction is in excess of its equilibrium value.

Many text books on solid state physics now use this result to reach immediately the current–voltage characteristic of the p–n junction rectifier. The argument is simply that we have two thermionic emitters facing one another and, as for the metal–semiconductor junction, we see that the current–voltage characteristic has the form of Equations (7.5) and (7.6), namely

$$I = I_0[\exp(eV_{ext}/k_BT) - 1] \tag{7.28a}$$

and, in reverse bias

$$I = I_0[1 - \exp(-eV_{\text{ext}}/k_B T)]. \tag{7.28b}$$

The only difference is that we now have two types of carrier, but clearly from the symmetry of Equation (7.27), the current–voltage characteristic for holes will be identical to that for electrons. Despite holes having opposite charge to electrons, the current flow is additive, since they flow in opposite directions.

However, this approach is not strictly correct for the p–n junction. The equation of the characteristic is indeed correct, as we shall see shortly, but implicit in the argument is the assumption that the junction is narrow. It is essential to assume that the probability of an electron recombining with a hole in the junction region is negligible. In most p–n junction devices this assumption is not valid and we must consider the diffusion of carriers across the junction in more detail. The effect of our more rigorous treatment is to alter the expression for the constant I_0.

Close to the junction we may consider the current to be entirely due to the diffusion current associated with carriers diffusing across the junction and we can neglect the current associated with the thermal generation of carriers in this region. As has been remarked previously, there is a probability that electrons and holes will recombine and we define the electron–hole recombination time τ_r as the average time which an electron in the conduction band remains free before recombining with a hole and returning to the valence band. As holes are involved in the recombination process the hole recombination time is the same as that for electrons. Close to the junction the rate of change with the time of the electron carrier concentration due to recombination will be proportional to the excess concentration n' above the concentration far from the junction region. If n_p is the concentration of electron carriers far from the junction and n is the concentration at a point in the junction region, then

$$\begin{aligned} -\partial n/\partial t &= (n - n_p)/\tau_r \\ &= n'/\tau_r. \end{aligned} \tag{7.29}$$

Consider now an element of the junction of thickness δx and of unit cross-sectional area. Then the number of electrons diffusing across this unit area per unit time at x is proportional to the concentration gradient with respect to distance. The number is $-D(\partial n/\partial x)$ where the proportionality constant D is referred to as the diffusion coefficient. At position $x + \delta x$ the number is $-D[(\partial n/\partial x) + (\partial^2 n/\partial x^2)\delta x]$.

The net difference is $-D(\partial^2 n/\partial x^2)\delta x$, and this loss of electrons must arise from the loss $(n'/\tau_r)\delta x$ due to recombination with holes in the thickness of junction δx. For a situation of dynamic equilibrium we must have

$$D(\partial^2 n/\partial x^2)\delta x = (n'/\tau_r)\delta x. \tag{7.30}$$

Equation (7.30) can be re-written as

$$D(\partial^2 n'/\partial x^2) = n'/\tau_r \tag{7.31}$$

and this has a solution of the form

$$n' = A\exp[-x/(D\tau_r)^{1/2}] + B\exp[x/(D\tau_r)^{1/2}]. \tag{7.32}$$

The coefficient B must be zero or the second term would 'blow up' at large distances from the junction, and hence

$$n' = n'(0)\exp(-x/L). \tag{7.33}$$

$n'(0)$ is the excess concentration of free electrons at $x = 0$ and $L = (D\tau_r)^{1/2}$, is known as the diffusion length. The excess concentration $n'(0)$ is related to $n(0)$ the concentration at $x = 0$ and n_p the concentration far from the junction in the p-side by

$$n'(0) = n_p - n(0). \tag{7.34}$$

Since $n(0)$ is equal to n_{p0} (the concentration in the p-region under zero bias) from Equation (7.27) we have

$$n'(0) = n_{p0}[\exp(eV_{ext}/k_B T) - 1]. \tag{7.35}$$

Combining Equations (7.33) and (7.35) gives

$$n' = n_{p0}\exp(-x/L)[\exp(eV_{ext}/k_B T) - 1]. \tag{7.36}$$

The diffusion current density J in the p-type region at $x = 0$ is

$$J(0) = -e(\partial n/\partial x) \tag{7.37}$$

and from (7.36) we obtain

$$J(0) = (eDn_{p0}/L)[\exp(eV_{ext}/k_B T) - 1]. \tag{7.38}$$

An identical analysis holds for holes and if we define electron and hole diffusion coefficients D_e and D_h and diffusion lengths L_e and L_h respectively, we have

$$J(0) = \left(\frac{eD_e n_{p0}}{L_e} + \frac{eD_h p_{n0}}{L_h}\right)[\exp(eV_{ext}/k_B T) - 1]. \tag{7.39}$$

This clearly has the same form as Equation (7.28). The current–voltage characteristic is shown in Fig. 7.2, and this gives quite good agreement with experimental characteristics for germanium diodes.

However, the experimental characteristic of a silicon diode does not look quite like Fig. 7.2. Although there is a very rapid rise of current in forward bias (Fig. 7.7) and rectifying action in reverse bias, we note that a forward bias of 0.6 volts is required before a forward current flows. This effect arises from the presence of traps at the junction, which substantially reduce the recombination time τ_r as they provide preferential recombination sites for the trapped electrons and holes. When the traps are predominantly empty at low current densities, their effect is so great as to inhibit all current flow, all carriers diffusing into the junction region recombining. Only when the traps are saturated at higher current densities does the forward current flow.

This effect has technical applications, as silicon diodes in forward bias acts as very high impedances to small signals but very low impedances to high voltages. A pair of silicon diodes, back to back between the input and earth is used to provide input over-voltage protection for sensitive high gain amplifiers and analogue to digital converters.

7.2.4 The Zener and avalanche diodes

Because the depletion layer is devoid of free carriers, it has very high impedance compared with the bulk n- and p-type material in the diode.

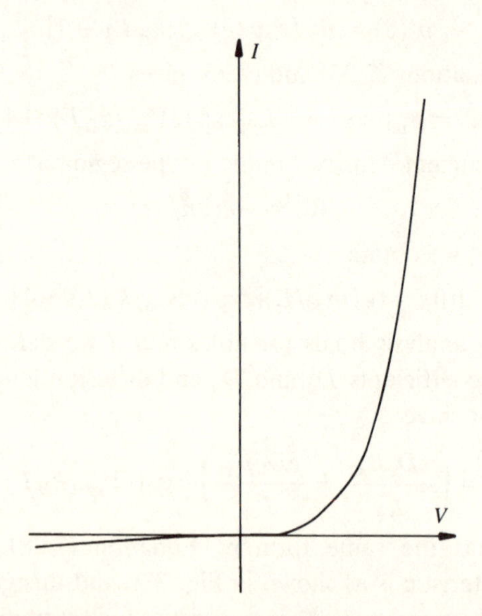

Fig. 7.7 $I-V$ characteristic of a p–n junction diode containing traps which lead to a non-zero turn on voltage.

This means that most of the voltage drop across the device in reverse bias occurs across the depletion layer. In a very heavily doped p–n junction diode, we have already seen that the depletion layer is very thin, and thus very intense electric fields are set up in this region. Electrons injected into the depletion region therefore experience very large accelerations and as a consequence gain a large amount of energy. At a critical energy these electrons have sufficient energy to ionize the atoms of the lattice, thereby releasing bonding electrons from the valence band into the conduction band. As these fresh carriers are injected into the conduction band, the current rises. The number of carriers thus created rises very rapidly with voltage and an avalanche process occurs, giving a very rapid rise in current for a very small change in reverse bias voltage (Fig. 7.8).

The applicability of this effect arises from the fact that it is a specific value of the electric field which determines the point at which the ionization process takes place. Therefore by tailoring the depletion layer thickness by varying the dopant concentration levels, the *voltage* at which the reverse breakdown occurs can be varied. It is thus possible to manufacture a range of diodes, known as Zener diodes, which break down under various values of reverse voltage. Such devices are extensively used as voltage regulators and voltage clamps as well as in protection circuits for delicate meters and components which will not stand a reverse voltage across them.

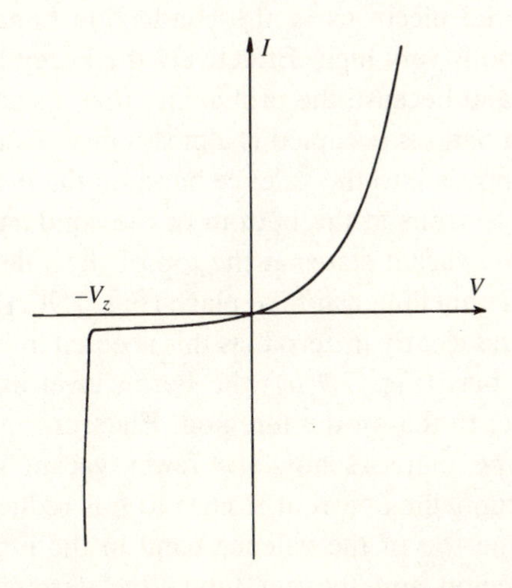

Fig. 7.8 I–V characteristic of a Zener diode. Breakdown occurs abruptly at reverse voltage $-V_z$.

Avalanche diodes operate in much the same way as Zener diodes and indeed there is often confusion over which process is operating in specific instances. In the avalanche process the free carriers, created by the ionization, themselves are accelerated sufficiently to cause ionization and an avalanche process results. Avalanche diodes are used as high gain photodetectors in which the electron–hole pairs created by a visible light or near infra-red photon set off the avalanche process and hence produce a large charge flow.

7.2.5 Tunnel diodes

The tunnel diode is another device which exploits very heavy doping to produce a very narrow junction region. While it is now effectively obsolete, it has a number of interesting features which make it worth our study at this point. Resonant tunnelling devices, which are somewhat analogous two-dimensional quantum well devices, are however attracting considerable attention at present. If the p- and n-material are extremely heavily doped, Equation (7.15) shows that the depletion region becomes very narrow. When this becomes only a few atomic spacings, there is a significant probability that electrons can quantum mechanically tunnel across the junction. This mode of charge transfer across the junction is very different from the diffusion mechanism just discussed.

When the doping is very heavy, say about 0.1%, the concentration of thermally excited electrons in the conduction band on the n-type side of the junction is very high. Effectively the Fermi level moves into the conduction band because the probability that a state at the bottom of the conduction band is occupied is almost unity. Similarly the Fermi level effectively moves into the valence band on the p-side. Then, with a thin junction, electrons at the bottom of the conduction band in the n-type material see vacant states at the top of the valence band in the p-type material. Tunnelling can take place (Fig. 7.9(*a*)) between states of equal energy and clearly in zero bias this is equal in both directions.

Under reverse bias (Fig. 7.9(*b*)) the Fermi level in the p-region is raised with respect to that in the n-region. Electrons in the conduction band in the n-type material now see fewer vacant states (at equal energy) and the tunnelling current from n to p is reduced. In contrast, the electrons at the top of the valence band in the p-region see more states in the n-region and the net tunnelling current from p- to n- increases rapidly.

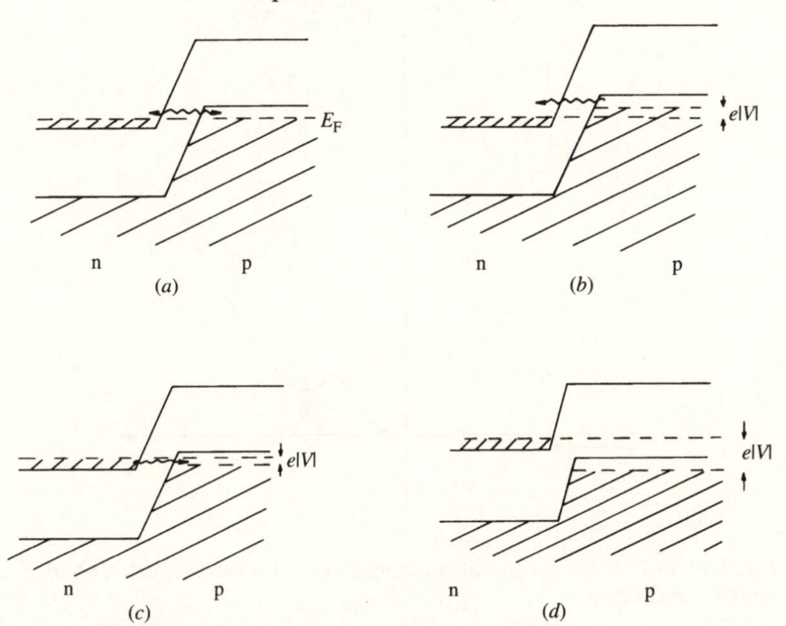

Fig. 7.9 Schematic diagram of the operation of a tunnel diode. (*a*) Zero bias (*b*) reverse bias (*c*) small forward bias (*d*) high forward bias.

It is, however, in forward bias that the tunnel diode exhibits its unique characteristic. As the bias is applied, electrons in the conduction band in the n-region see more states available in the p-region (Fig. 7.9(*c*)) and the current from n to p increases. Conversely, the p to n current is decreased and the net tunnelling current increases. It does not increase indefinitely because the number of empty, equal energy states seen by the n-region conduction electrons reaches a maximum when the Fermi level on the n-side is level with the top of the valence band on the p-side. With further increase in bias, the density of equivalent states on either side decreases, electrons in the n-type side see no states of the same energy in the p-side. Tunnelling then stops (Fig. 7.9(*d*)) and the diode continues to operate as a normal p–n junction diode in forward bias. As a result the characteristic has the form shown in Fig. 7.10. Between points *D* and *E* where the density of equivalent states on the two sides is falling, the device displays a *negative resistance* region. A device which has an increasing current for a decreasing voltage will, when the output is fed back to the input, oscillate. In the case of the tunnel diode this oscillation can be of very high frequency and microwaves are emitted. Tunnel diodes were

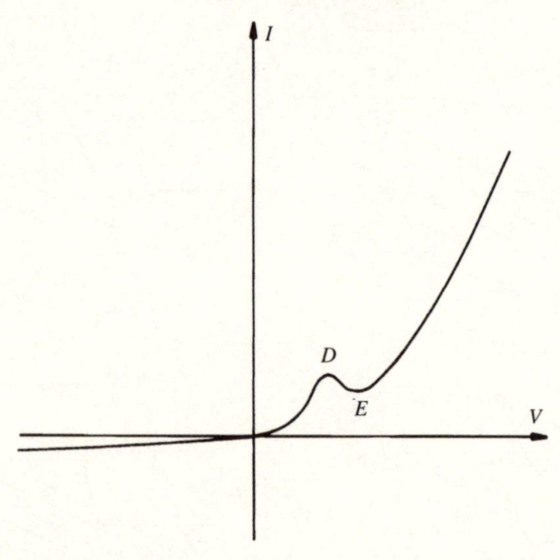

Fig. 7.10 $I-V$ characteristic of a tunnel diode. Region between D and E has negative resistance.

difficult to manufacture reliably and they have now been almost totally superseded by the Gunn devices which rely on the transfer of electrons or holes between energy bands where the effective masses are very different.

7.3 The bipolar transistor

An n–p–n transistor essentially consists of two p–n junction diodes back to back, the p-region being very thin. In equilibrium, two depletion layers are formed, as shown in Fig. 7.11(a). By convention, the left hand region is called the *emitter*, the central region the *base* and the right hand region the *collector*. If we forward bias the base with respect to the emitter, electrons in the emitter flow easily into the base. In the base they become minority carriers and if the base region is thick, they will recombine with holes. However, if the base is thin compared with the average distance that a minority carrier travels before recombination (about 1 mm) and the collector is reverse biased with respect to the base (Fig. 7.11(b)), then the electrons are swept into the collector. For a very thin base region, virtually all the electrons released from the emitter reach the collector. In terms of the circuit shown in Fig. 7.11(c) the emitter and collector currents are

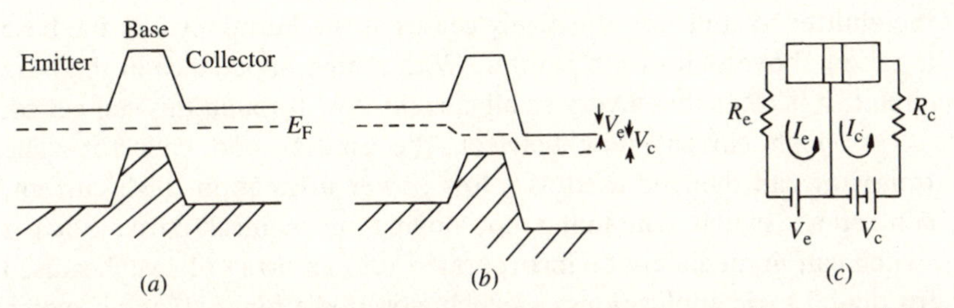

Fig. 7.11 (*a*) Band structure of an n–p–n transistor in zero bias. (*b*) Band structure with base (b) forward biased with respect to the emitter (e) and collector (c) forward biased with respect to base. (*c*) Equivalent circuit diagram to the situation in (*b*).

equal, i.e. $I_e = I_c$. The voltage V_c developed across the output resistor R_c is $I_c R_c$ and similarly the input voltage V_e is $I_e R_e$. Therefore we have

$$V_c/V_e = R_c/R_e. \tag{7.40}$$

Because the junction impedances are very high, we can make $R_c \gg R_e$ and we can obtain voltage amplification by transfer of charge from a low impedance circuit to a high impedance one. Because the power dissipation in the emitter and collector circuits is $I_e^2 R_e$ and $I_c^2 R_c$ respectively, it is also seen that power amplification occurs in the same ratio as the voltage amplification. This transfer process led to the original name transfer-varistor for the device which has subsequently been shortened to transistor as the term varistor ceased to be used for the p–n diode.

An important feature of the transistor is the ability to use it as a switch. In the circuit in Fig. 7.12, the bias on the base is changed. As the base potential is raised, it is clear that the flow of electrons from

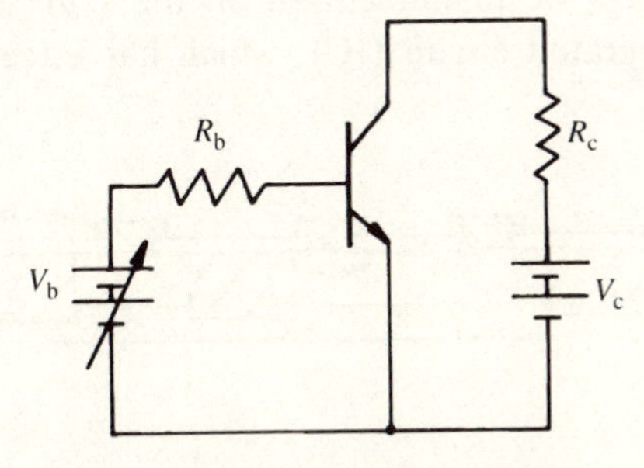

Fig. 7.12 Circuit for use of a bipolar transistor as a switch.

the emitter to collector effectively ceases as the Fermi level of the base is raised above that of the emitter. With a high impedance in the base circuit, it is clear that a very small current flow through this can cut off a very high current flow between the emitter and collector. The transistor can then be used as a low power dissipation, high current, contactless switch containing no moving mechanical parts. Such a switch can immediately be incorporated into an array of switches used for digital logic applications; a switch closed is a binary 0 and a switch open is a binary 1.

Transistors are now made by a process of *planar technology* in which many devices are fabricated simultaneously on one thin wafer of silicon, typically 0.15 mm thick. A schematic diagram of the simplest form of bipolar transistor is shown in Fig. 7.13. The polished n-type wafer is oxidized in steam at a high temperature to produce a thin layer of SiO_2 on the surface. This is coated with a photoresistive substance which is exposed to light in the area associated with the transistor base region. The photoresist can be developed in chemicals in the regions exposed to light, opening a window onto the SiO_2 surface. This is removed by etching in HF, which will attack SiO_2 but not Si, leaving a window through to the silicon itself. Through this is diffused p-type impurity, e.g. boron from a boric oxide glass deposited on the surface and driven in by a high temperature heat treatment. The steps are repeated to build up the structure shown in Fig. 7.13. (P_2O_5 is diffused into the p-type region below the emitter (e) to yield n-type material.) Metal contacts, usually aluminium, are then added to provide contacts. Many such devices can be manufactured at the same time on one 5 or 6 inch silicon wafer, and these are separated by scribing with a diamond stylus and fracturing, in a manner extensively used on a larger scale by plate glass-cutters.

The whole process has been miniaturized and arrays of planar transistors can now be manufactured on the same wafer. This is the basis of the integrated circuit (IC), which has effectively created the

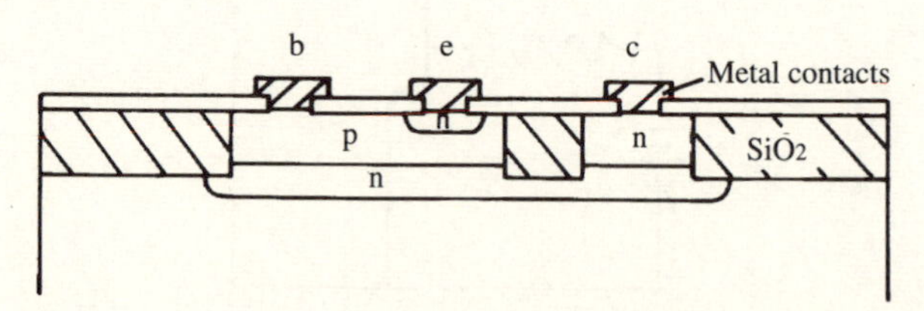

Fig. 7.13 Structure of a planar bipolar transistor fabricated by IC technology.

field of information technology. Here is not the place to discuss IC technology and the reader is referred to electronic engineering books such as D. V. Morgan and K. Board, *An Introduction to Semiconductor Microtechnology* (Wiley: 1983).

7.3.1 The field effect transistor

The field effect transistor works in a somewhat different manner. We begin by considering the n-channel junction field effect transistor (JFET), shown schematically in Fig. 7.14. The device, which is amenable to fabrication by planar technology, operates by the control of the flow of electrons in the n-type channel region by variation of the voltage applied to the p-type gate region. Under zero bias field, a depletion region is set up between the n- and p-regions as shown in the cross-hatched region of Fig. 7.14(*a*). Because the device is fabricated with the p-region being much more heavily doped than the n-type regions, the depletion region extends into the n-type channel region below the gate. If the drain is forward biased with respect to the source (drain positive), and the gate is reverse biased with respect to the source (gate negative) then the depletion region expands as shown in Fig. 7.14(*b*). Thus an increase of the reverse biased gate voltage reduces the cross-sectional area of the channel and the resistance

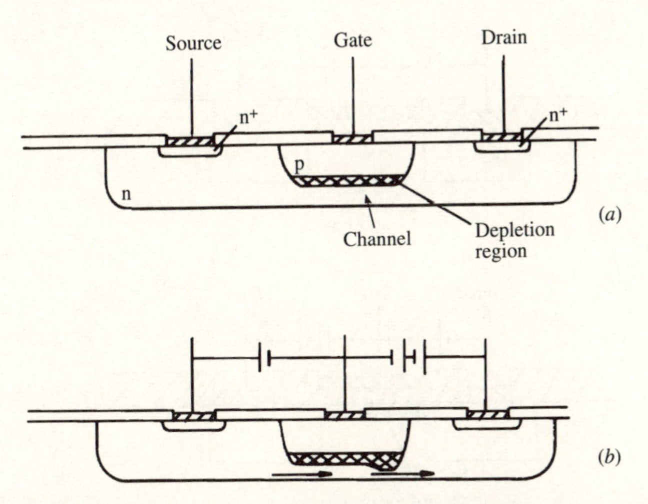

Fig. 7.14 Structure of a junction field effect transistor (JFET). Regions marked n and n^+ have different impurity concentrations. (*a*) Zero bias (*b*) drain biased forward with respect to the source and gate reverse biased with respect to the source.

increases, reducing the current flowing. Further increase in the gate voltage eventually results in the depletion region extending right through the channel and 'pinching-off' the current flow. Thus, like the bipolar transistor, the JFET can be used as a highly efficient switch.

While the bipolar transistor is extremely fast and in appropriate form is capable of handling high currents it has several processing stages and is therefore (relatively) expensive to manufacture. The metal oxide semiconductor (MOS) field effect transistor has fewer processing stages and has come to dominate the large arrays of devices used in ICs for logic purposes. A schematic diagram is shown in Fig. 7.15, where we see that the p-type gate region is now missing and the source and drain are separated. Current flow occurs through a channel just below the oxide surface and is controlled by the voltage on a metal electrode, insulated from the channel by the thin film of SiO_2. Power consumption is thus very low. The channel region occurs because, close to the metal–oxide interface, the bands are bent as in the case of the metal–semiconductor junction discussed earlier. This results in an inversion layer of electrons when in forward bias which can be pinched-off by reversal of the gate voltage. Use of complementary n-channel and p-channel devices in pairs makes it possible to design logic circuits where current only flows when switching is actually taking place and thus power consumption can be reduced dramatically.

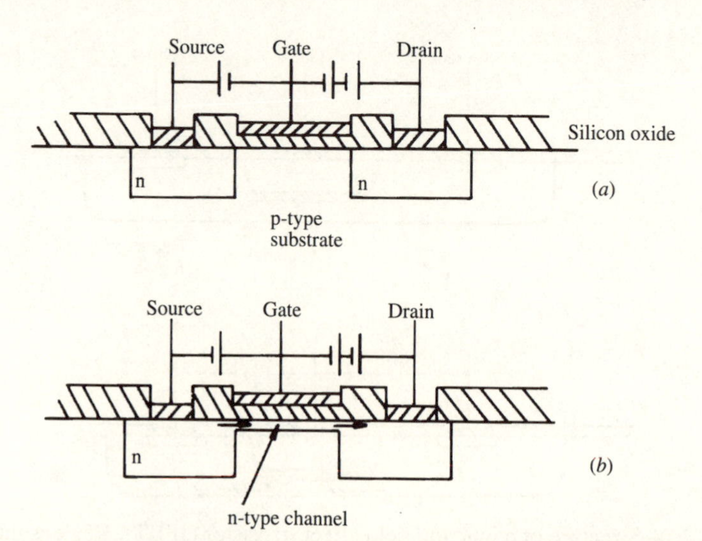

Fig. 7.15 Structure of metal oxide semiconductor field effect transistor (MOSFET). (*a*) Gate forward biased with respect to its source. Drain forward biased with respect to gate. (*b*) Gate reverse biased with respect to its source.

7.4 Electro-optic devices

Light emitting devices are becoming of increasing importance, not only as display devices but also as means of sensing recorded information as in a compact disc player. A rapidly developing area is in use of such devices for transmission of data at a very high rate through optical fibres, either in electrically noisy environments or over large distances in long-haul telecommunications links. All rely on the same principle, namely that electron and hole recombination takes place with emission of a photon of well defined frequency. For this to happen the material must have a direct gap, that is the minimum in the conduction band must be directly above the maximum in the valence band of the $E-k$ diagram (Fig. 7.16(*a*)). The III–V compounds such as GaAs and InP have such band structures, unlike Si and Ge. Electrons and holes excited in the conduction and valence bands respectively will thermalize rapidly and occupy states close to the band edges. Thus the most probable recombination of electrons and holes occurs between such states and the energy released is the energy of the band gap E_g. Typical photon wavevectors are 10^5 cm^{-1} and thus in the context of the width of the first Brillouin zone (typically 10^8 cm^{-1}) the transition is effectively vertical. Conservation of energy and momentum are both possible as $\hbar \Delta k \approx 0$. This is not the case for an indirect gap semiconductor (Fig. 7.16(*b*)). Here the electron and hole do not have equal wavevectors and therefore there is momentum to carry away. If the conduction band minimum is at the zone boundary this will typically be

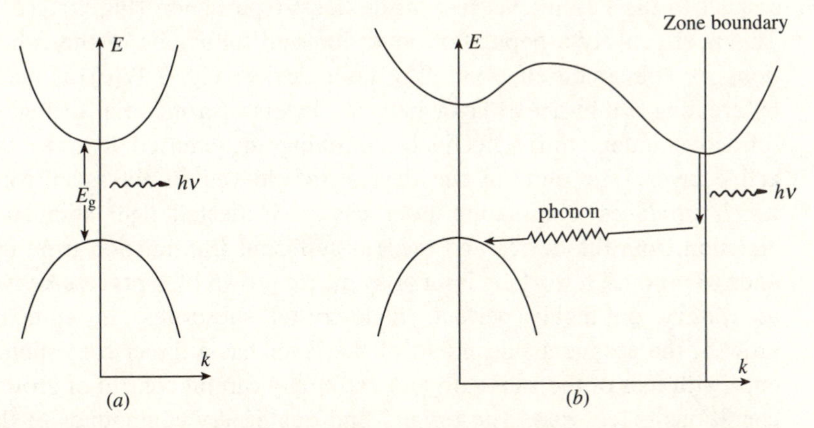

Fig. 7.16 (*a*) $E-k$ diagram of a direct gap semiconductor showing vertical transition involving electron and hole recombination with sole emission of a photon. (*b*) An indirect gap semiconductor showing recombination with emission of both a phonon and a photon.

$\hbar\pi/a$ where a is a lattice spacing. The only way to conserve both energy and momentum in the recombination process is for a phonon, a quantum of crystal lattice vibration, to carry away a small part of the energy but almost all of the momentum. As a result, there is a wide distribution of energies of the photons emitted and indirect gap materials such as Si and Ge cannot be used for electro-optic applications.

A forward biased p–n junction in a direct gap material such as GaAs will emit monochromatic radiation. Electrons in the n-type region diffuse across the depletion region where they recombine with holes in a narrow region just inside the p-material. The frequency of the light emitted is determined by the energy gap. For GaAs this is 0.87 μm, in the near infra-red while for GaP this is 0.55 μm, in the red region of the spectrum. By appropriate alloying of the Group III (Ga, In, Al) and Group V (As, P, Sb) elements in ternary compounds such as $GaAs_xP_{1-x}$ or quaternary compounds such as $Ga_xIn_{1-x}As_yP_{1-y}$, the band gap can be tailored to give emission over a wide range. Thus red, green and yellow light emitting diodes can be fabricated.

7.4.1 The solid state injection laser

When the n-region of the structure shown in Fig. 7.17(*a*) is very highly doped, the Fermi level may enter the conduction band. Under sufficiently high forward bias, the electrons 'spill over' into the p-type region creating a very high density of electrons in high energy states with respect to the Fermi level just inside the p-type region (Fig. 7.17(*b*)). This is effectively a population inversion and fulfils one of the conditions for stimulated emission. The laser device (Fig. 7.17(*c*)) is made by creating a thin active layer between layers of a material of higher refractive index, thus effectively confining the emitted light to the active layer. The ends of the device are cleaved to act as mirrors, which produces a miniature laser cavity. Reflected light stimulates emission from the device and there is sufficient transmission from the ends to produce a working laser. Layers are grown by a process known as *epitaxy* on highly perfect single-crystal substrates. In epitaxial growth, the atomic arrangement of the layer has a direct correspondence with that of the substrate and extremely careful control of growth conditions is required. The ternary and quaternary compounds of the Group III and V elements used have the same crystal structure as GaAs and InP but the crystal lattice spacing varies with composition. As a result the layer is strained with respect to the substrate. The mismatch in lattice parameter must be kept small in order that the layer does not relax by introduction of dislocations at the interface. The InGaAsP lasers for telecommunications applications are grown with a composition such that the emitted light matches the minimum in the absorption of the glasses used for fibre-optics. Lasers for compact disc players are usually made from an active layer of $Ga_xAl_{1-x}As$ on GaAs.

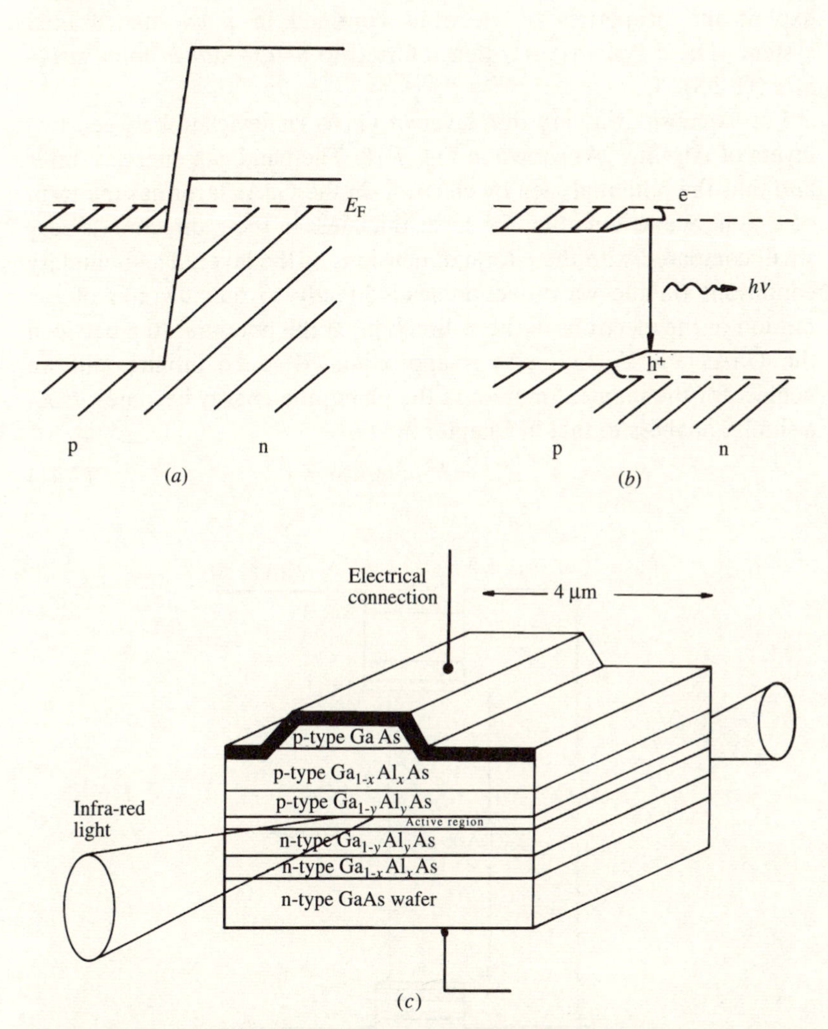

Fig. 7.17 The injection laser. (*a*) Band structure of highly doped p–n junction with Fermi levels in conduction and valence bands. (*b*) Creation of a population inversion under forward bias and recombination in the junction region. (*c*) Structure of a laser made by epitaxial techniques.

7.5 Quantum well devices

The single layer laser has a disadvantage that the threshold current densities are high with consequent problems of reliability. By making the laser from a stack of very thin active layers separated by thin confining layers, the thresholds can be dramatically reduced. The multi-layer laser can also be tuned in wavelength by adjustment of the layer thickness and is just one member of a class of novel devices which exploit the properties of electrons confined in a two-dimensional system. These systems are often referred to as *low dimensional structures* (LDS).

Let us consider a very thin layer of GaAs sandwiched between two layers of $Al_xGa_{1-x}As$ shown in Fig. 7.18. The band gap energies differ and thus the potential seen by electrons in the GaAs layer has the form of a well. We assume that the layer thickness in the z direction is very small compared with the lateral dimensions of the layer. The boundary conditions on the wavefunction lead directly to quantization of the motion of the electrons in the z direction. If the potential step between the GaAs and $Al_xGa_{1-x}As$ is approximated to be infinite and we neglect for the moment motion in the plane, the energy becomes, from a similar analysis to that in Chapter 2,

$$E_{cz} = h^2n^2/8m_e^*d^2 \tag{7.41}$$

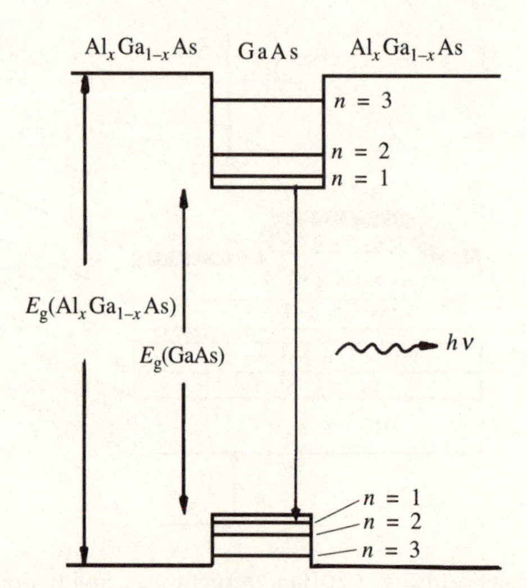

Fig. 7.18 GaAs quantum well. Thin GaAs layer sandwiched between two $Al_xGa_{1-x}As$ layers leads to discrete well levels and tunable light emission from states in the quantum well.

where d is the GaAs thickness, $n = 1, 2, 3, \ldots$ etc. and the zero is taken as the bottom of the conduction band in bulk GaAs. As the motion is effectively unrestricted in the layer plane, the total energy becomes, assuming parabolic bands,

$$E_c = \hbar^2 k_{\|}^2/2m_e^* + h^2 n^2/8m_e^* d^2. \tag{7.42}$$

Where $k_{\|}$ is the component of the wavevector in the plane of the film, and m_e^* and m_h^* are the electron and hole effective masses. Similarly for the states in the valence band, we have

$$E_v = \hbar^2 k_{\|}^2/2m_h^* - E_g - h^2 n^2/8m_h^* d^2. \tag{7.43}$$

The effect of a finite size barrier is to modify the absolute values of the energy levels, as the z component of the wavefunction becomes non-zero at the edge of the well (Fig. 7.19(*a*)). The $E-k$ curves then have the form shown in Fig. 7.19(*b*), with corresponding steps in the density of states (Fig. 7.19(*c*)).

7.5.1 Quantum well lasers

Let us now recall the process of emission due to recombination of electrons and holes in a direct gap material. The lowest conduction and highest valence band energies occur when $k_{\|} = 0$ and thus the frequency of the emitted light ν from a quantum well is given by

$$h\nu = E_g + h^2/8m_e^* d^2 + h^2/8m_h^* d^2. \tag{7.44}$$

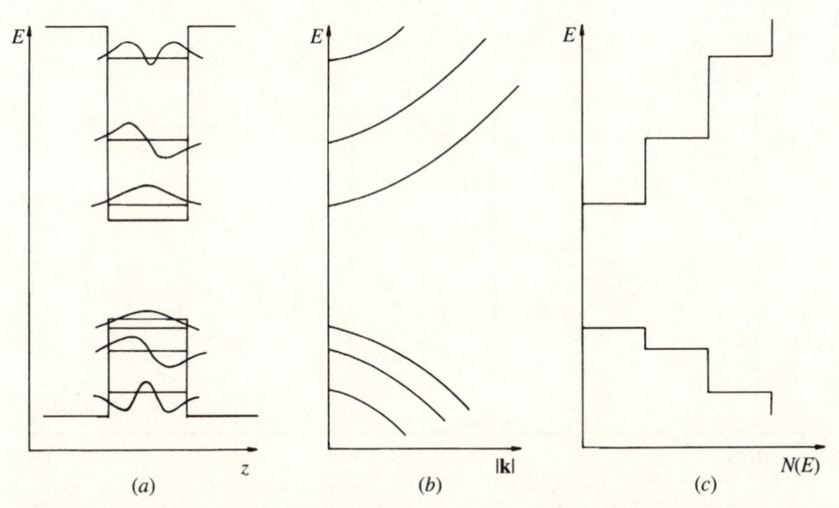

Fig. 7.19 (*a*) Wavefunctions associated with the well states. (*b*) $E-k$ curves including in-plane component of wavevector. (*c*) Density of states as a function of energy.

Thus we see that the emission frequency is strongly controlled by the well width and thus a tunable electro-optic device can be made. Incorporation of a single quantum well into a laser is less common than use of several such layers in an alternating sequence, as this enables the threshold current density to be reduced. For a periodic sequence, Fig. 7.20(*a*), in which there is significant wavefunction penetration between wells there is periodicity of $2\pi/d$ in k space, rather than $2\pi/a$ and the band is split up into sub-bands within a mini-Brillouin zone between $k = \pm\pi/d$ (Fig. 7.20(*b*)). The energies at $k_{\parallel} = 0$ are again shifted. Such a sequence is known as a superlattice, or multiquantum well if the barriers are sufficiently thin to allow significant penetration of the wavefunctions between wells.

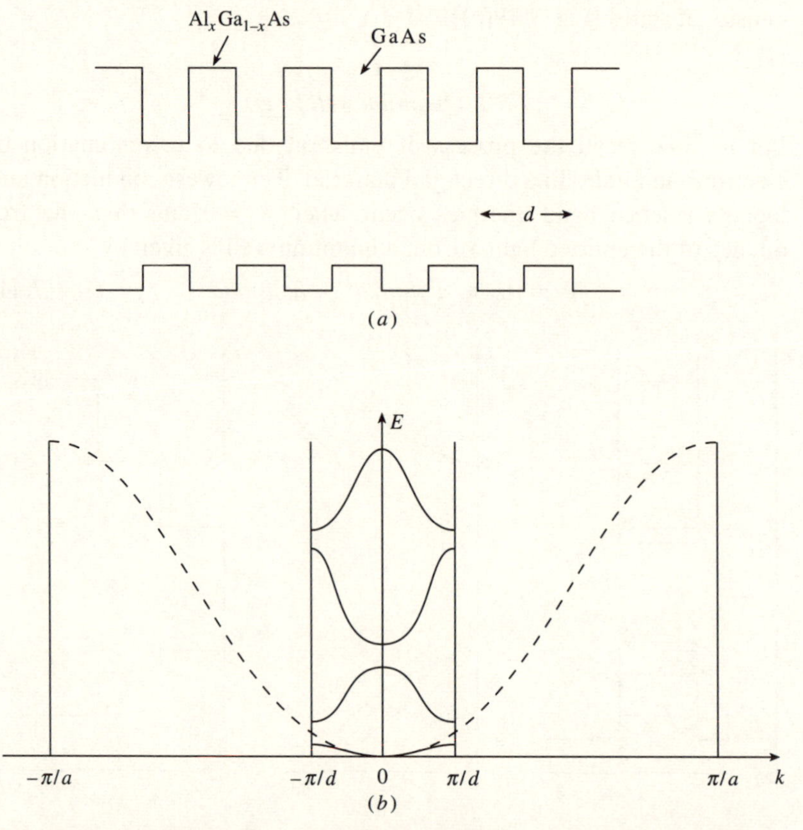

Fig. 7.20 (*a*) Multiquantum well (MQW) structure with significant wavefunction overlap between wells. (*b*) Band diagram showing the mini-Brillouin zone arising from the periodicity of the MQW structure.

7.5.2 High electron mobility devices

If the $Al_xGa_{1-x}As$ barrier layers are doped with n-type impurity, electrons are transferred from the n-type layers to the undoped wells as a process of equalizing the Fermi levels. The resulting space charge results in bending of the bands. As the electrons predominantly occupy the lowest energy levels, a separation occurs between the donor electrons and their host impurity atoms. The electrons are therefore not scattered by the ionized impurities and have a very substantially increased mobility. High electron mobility transistors (HEMTs) based on GaAs quantum well devices are now in production and are important for microwave and radar applications as well as being candidates for devices in the next generation of computers. The physics of quantum well devices is presently an extremely active area and a whole range of new devices has been proposed, many of which are gradually being realized as growth and fabrication techniques continue to improve.

Problems

7.1 Show that the capacitance of a p–n junction is the same as that of a parallel-plate capacitor with plate separation equal to the width of the depletion layers and filled with dielectric of the same value as that of the semiconductor.

The energy gap in Ge is 0.66 eV at room temperature, and the dielectric constant is 16. A junction is made between n-type material with 10^{21} donor atoms per cubic metre and p-type material with 10^{21} acceptor atoms per cubic metre. If the junction area is 1 mm^2, determine the total depletion region width and the junction capacitance in (*a*) zero applied voltage and (*b*) an applied voltage of 5 V.

7.2 In a particular application, there is a potential drop of 0.1 V across an ideal diode. If the maximum ratio of reverse to forward current permitted is 10^{-5}, determine the maximum operating temperature of the diode.

7.3 In a particular transistor, the product *np* at a temperature of 300 K is 6.3×10^{16} times that at 100 K. Determine the value of the energy gap.

8
Localized electrons

In this chapter we examine the way in which the simple model of an electron in a box can be applied to understand the optical properties of the alkali halides. We then proceed to consider the magnetic properties of electrons which are localized at specific lattice sites. With this discussion of the phenomena of diamagnetism and paramagnetism, we lead up to the next chapter on magnetic order where we see the breakdown of the independent electron approximation giving quite dramatic results.

8.1 Point defects in alkali halides

The band gaps of the alkali halides, such as NaCl or KCl, are high and they are found to be excellent insulators. Thus, in the pure and perfect state, the alkali halides are transparent to visible light. However, as a result of either irradiation with X-rays, heating in the vapour of the alkali metal or electrolysis at high temperature, it is possible to create large numbers of negative ion vacancies (Fig. 8.1). These defects correspond to unoccupied lattice sites which would normally contain a halide ion. If the lattice site was simply vacant, the site would appear to be positively charged due to the removal of negative charge on the halide ion. In order to preserve electrical neutrality, it is favourable for an electron to become trapped at this negative ion vacancy. Such an electron trapped at a negative ion vacancy is called an F or colour centre. (The term F centre stems from the German word for colour– 'fabre'.) As we will see below, such defects give rise to intense and characteristic colouration of the crystals.

Let us attempt to model the optical properties of the F centre by assuming that the electron is confined to a cubic box of infinitely deep

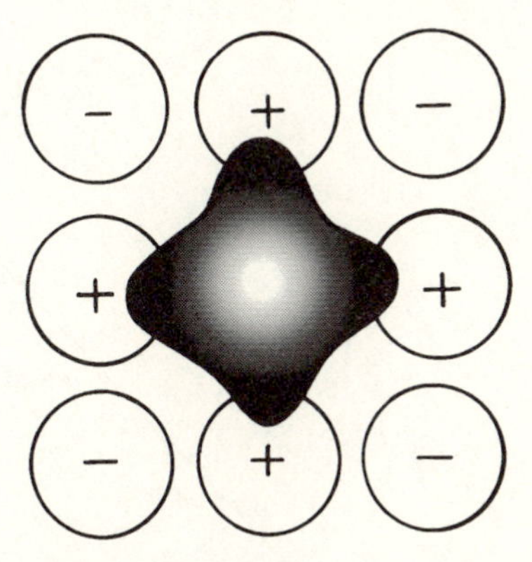

Fig. 8.1 Schematic diagram of a negative ion vacancy showing a trapped electron at an unoccupied lattice site.

potential. The alkali halides have a face centred cubic structure with a basis of an alkali ion at [000] and a halide ion at $[0\frac{1}{2}\frac{1}{2}]$. We see from Fig 8.1 that the length of the box confining the electron is thus equal to the spacing of the alkali ions along the cube edges divided by $\sqrt{2}$. We can now solve the three-dimensional Schrödinger equation exactly as we did in Chapter 2. From Equation (2.32) we see that the energies of the trapped electron E_n are quantized and given by

$$E_n = (\hbar^2\pi^2/ma^2)(n_x^2 + n_y^2 + n_z^2) \tag{8.1}$$

where n_x, n_y and n_z are integers and a is the lattice parameter. The lowest energy state is when $n_x = 0$, $n_y = 0$, $n_z = 1$, and thus the ground state energy E_0 is given by

$$E_0 = (h^2/4ma^2). \tag{8.2}$$

The first excited state is when $n_x = 0$, $n_y = 1$, $n_2 = 1$. Thus

$$E_1 = (h^2/2ma^2). \tag{8.3}$$

When light is incident on the alkali halide crystal containing these F centres, strong absorption occurs when the photon energy is equal to the energy difference between the ground state and first excited state. The photon energy is used to excite the trapped electron and an absorption band is found centred on frequency, ν_0, given by

$$\nu_0 = (h/4ma^2). \tag{8.4}$$

Experimental values for the energy at which the absorption maximum occurs are given in Fig. 8.2 and compared with the value

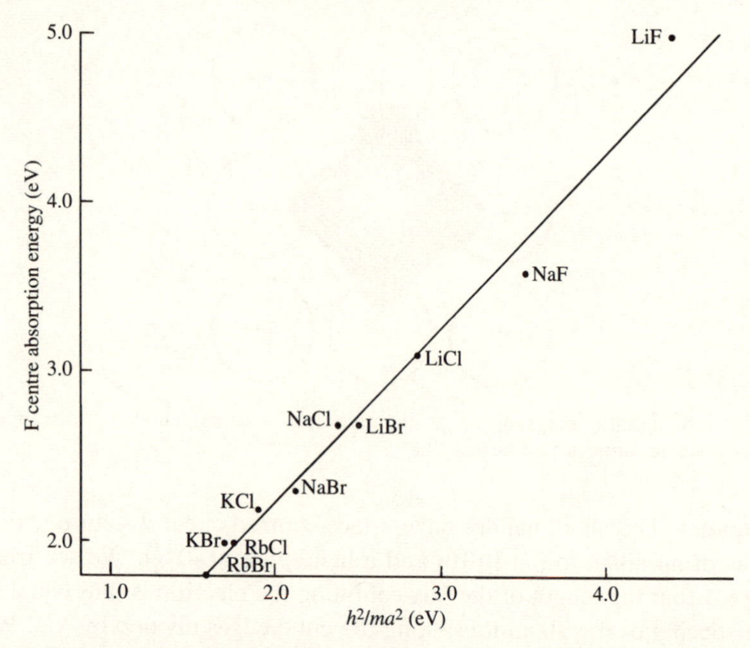

Fig. 8.2 Variation of the F band absorption energy maximum with the lattice spacing a of a range of alkali halides. Unit gradient shows good agreement with the a^{-2} variation predicted by Equation (8.4).

evaluated using Equation (8.4). For such a simple model the agreement is very good. Due to this F band absorption. NaCl crystals containing F centres appear reddish-brown, LiF crystals appear light blue and KCl crystals display a magnificent deep-blue/purple colour. The F band absorption maximum varies inversely as the square of the crystal lattice parameter and this relation is found to hold extremely well through the whole range of alkali halides (Fig. 8.2).

8.1.1 Electron spin resonance of colour centres

The behaviour of point defects in alkali halides is extremely complex and due to association of adjacent defects more than one type of colour centre can be produced, giving rise to a variety of absorption bands. A number of techniques have been used to study this complex situation including thermoluminescence (Chapter 6). By heating the material, the electrons trapped at the colour centres can be released and detected from the ultra-violet and visible light they emit when they recombine with holes or positively charged acceptor sites. The amount of thermoluminescence emitted from such defects is proportional to the defect density and therefore to the X- and gamma-ray irradiation dose. Alkali halide TL dosimeters are used to monitor radiation doses received by people who are likely to be exposed to ionizing radiation in the course of their work.

Thermoluminescence provides a measure of the defect density but no details of the microscopic structure of the colour centre. One of the most powerful techniques for such studies is electron spin resonance (ESR). We have already seen in Chapter 3 that the electron possesses a magnetic dipole moment and that in a magnetic field B the moment can be either oriented parallel or antiparallel to the field. Unlike the classical one, there can only be two possibilities, the energy difference between the two states being given by δE where

$$\delta E = 2\mu_B B \tag{8.5}$$

(Fig. 3.3) where μ_B is the Bohr magneton. Magnetic resonance spectroscopy is the term given to the study of states whose energies are a function of magnetic field and the splitting of energy levels is tuned by slowly sweeping a quasi-static d.c. magnetic field. The energy level splitting of the electron trapped at an F centre is an excellent example. In an ESR experiment, the sample is placed inside a waveguide cavity or a tuned coil and bathed in a microwave or radio frequency (r.f.) field of constant frequency ω. This is situated in a quasi-static magnetic field which is slowly swept (Fig. 8.3(a)). Power is absorbed from the

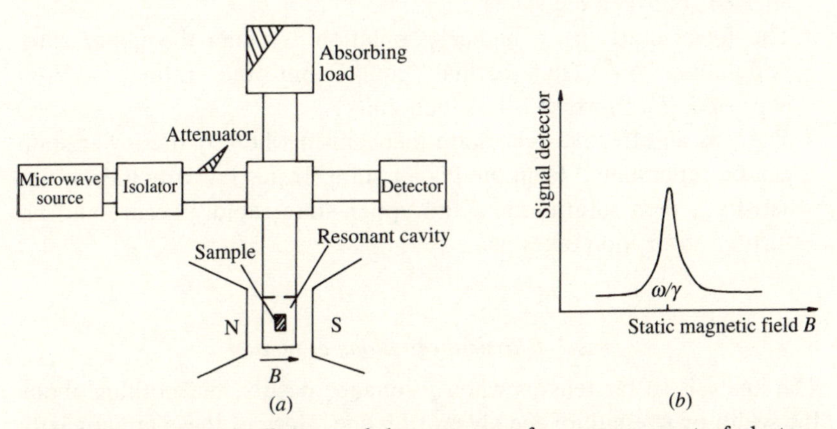

Fig. 8.3 (a) Schematic diagram of the apparatus for measurement of electron spin resonance. (b) Schematic diagram of detector signal as quasi-static field is swept through the resonance condition; γ is the gyromagnetic ratio.

microwave or r.f. field when the energy δE and the frequency ω are related by

$$\delta E = 2\mu_B B = \hbar\omega \tag{8.6}$$

and the resonance is thus detected (Fig. 8.3(*b*)). In practice, the sweep field is modulated at a low frequency to permit the use of phase sensitive detectors. Choice of frequency is usually based on convenience though the absorbed power rises with frequency. Use of the standard 3 cm microwaves for ESR gives resonance fields of about 0.3 tesla, which are quite easy to achieve with a high level of uniformity in the laboratory. Measurement of the ESR gives a value of the g factor (Equation (3.16)) for F centres very close to 2, showing that the model of a single electron trapped in a relatively large potential well is sound. When measured with high resolution, the resonance lines are seen to be split into many components which arise from the interaction of the electron with the field due to the surrounding ions and nuclei.

8.2 General principles of magnetic resonance experiments

All magnetic resonance techniques exploit the fact that energy can be absorbed in a system which has two energy states if:

1 The selection rules are correct and the quantum numbers of the two states are such that a magnetic dipole transition is allowed between them
2 The incident radiation has frequency $h\omega$ equal to the energy difference δE between the states
3 The lower state has a higher population N_1 than the upper state population N_2. (In thermal equilibrium the ratio $N_2/N_1 = \exp(-\delta E/k_B T)$ which is less than unity)
4 There is an effective relaxation mechanism whereby the lower state can be repopulated from the higher state. If this last condition is not satisfied, then saturation of the upper state rapidly occurs and no further absorption takes place.

8.2.1 Classical equations of motion

The analysis so far tells us when resonance occurs, but nothing about the width or strength of the absorption line. Both of these are crucially important in practice. They are determined by the relaxation time, or the loss mechanisms whereby equilibrium is re-established.

Consider the electron spin as a classical moment $\boldsymbol{\mu}$ which precesses in a uniform (static) field. The torque is $\boldsymbol{\mu} \times \mathbf{B}$ and equal to the rate of change of angular momentum. Thus

$$\hbar \mathrm{d}\mathbf{S}/\mathrm{d}t = \boldsymbol{\mu} \times \mathbf{B} \tag{8.7}$$

or, as

$$\boldsymbol{\mu} = g_s \mu_B \mathbf{S}, \tag{8.8}$$

$$\mathrm{d}\mu/\mathrm{dt} = \gamma(\boldsymbol{\mu} \times \mathbf{B}) \tag{8.9}$$

where $\gamma = g_s \mu_B/\hbar$ is the gyromagnetic ratio.

Thus the magnetic moment per unit volume $\mathbf{M}$ is given by

$$\mathrm{d}\mathbf{M}/\mathrm{d}t = \gamma(\mathbf{M} \times \mathbf{B}). \tag{8.10}$$

This is very general formula applicable to all resonance experiments. If the component of $\mathbf{M}$ parallel to the field M_z is not in thermal equilibrium, it relaxes towards its equilibrium value M_0 with relaxation time T_1 such that

$$\mathrm{d}M_z/\mathrm{d}t = (M_0 - M_z)/T_1. \tag{8.11}$$

Solution of this of course leads to an exponential decay.

Generally we can write the Bloch equations

$$\mathrm{d}M_z/\mathrm{d}t = \gamma(\mathbf{M} \times \mathbf{B})_z + (M_0 - M_z)/T_1 \tag{8.12a}$$

$$\mathrm{d}M_y/\mathrm{d}t = \gamma(\mathbf{M} \times \mathbf{B})_y - M_y/T_2 \tag{8.12b}$$

$$\mathrm{d}M_x/\mathrm{d}t = \gamma(\mathbf{M} \times \mathbf{B})_x - M_x/T_2. \tag{8.12c}$$

T_2 is the transverse relaxation time. This is very different from T_1 and is essentially a dephasing time of the precessional motion as there is no energy required to disorder the spins in the x or y directions. Solution of the Bloch equations for a static field in the z direction and a rotating field in a plane normal to z gives for the absorbed power

$$P(\omega) = (\omega\gamma M_z T_2 B_i^2)/[1 + (\omega_0 - \omega)^2 T_2^2] \tag{8.13}$$

and the half width at half the maximum power is

$$\Delta\omega = 1/T_2 = \gamma B_i. \tag{8.14}$$

Thus for a short relaxation time, the resonance peak is very broad and this is often the case in metals. Note that T_1 does not enter the relation for line width, although it affects total power absorbed. Normally $T_1 \gg T_2$. The term T_1 involves energy loss and is usually a measure of the spin–lattice coupling strength. Possible relaxation channels are

1 Phonon emission
2 Raman (inelastic) scattering of phonons
3 Two stage phonon processes.

8.3 Paramagnetism

When the magnetic moment of an alkali halide is measured as a function of magnetic field, behaviour is found to depend on whether or not the crystal contains colour centres. The moment associated with the colour centres is proportional to the applied magnetic field and in the same sense. Clearly the magnetic field is aligning the moments of the electrons localized at the negative ion vacancy sites.

8.3.1 Case of zero orbital angular momentum

More generally, we find this phenomenon occurs when unpaired electrons are localized at crystal lattice sites. Suppose that in the crystal there are N atoms each with one unpaired electron with spin $S = \frac{1}{2}$ and there are no interactions between these electrons. This implies that the electrons under consideration have no orbital angular momentum, a point we will return to later. Then, in a magnetic field B_0 the electronic ground state is split by an amount $g_s\mu_B B_0$, or as $g_s = 2$, by $2\mu_B B_0$. The lower level corresponds to $m_s = -\frac{1}{2}$ and the upper level to $m_s = +\frac{1}{2}$, where m_s is the quantum number corresponding to the z component of electron spin. The magnetic moment of the upper level is $-\mu$, i.e. antiparallel to the field while the lower level is $+\mu$, i.e. parallel to the field.

Let the population of the lower level be N_1 and that of the upper level be N_2. Then, if no other higher energy states are excited, clearly

$$N = N_1 + N_2. \tag{8.15}$$

All the electrons are localized at crystal lattice sites, they are in principle distinguishable, and therefore we may use Boltzmann statistics to determine the relative equilibrium populations of the two levels. Hence,

$$N_2/N_1 = \exp(-2\mu_B B_0/k_B T) \tag{8.16}$$

and using Equation (8.15), we see

$$\begin{aligned} N_1/N &= 1/[1 + \exp(-2\mu_B B_0/k_B T)] \\ &= [\exp(\mu_B B_0/k_B T)]/[\exp(\mu_B B_0/k_B T) + \exp(-\mu_B B_0/k_B T)] \end{aligned} \tag{8.17a}$$

and

$$N_2/N = [\exp(-\mu_B B_0/k_B T)]/[\exp(\mu_B B_0/k_B T) + \exp(-\mu_B B_0/k_B T)]. \tag{8.17b}$$

The net magnetic moment M in the direction parallel to the field is given by

$$M = N\langle\mu\rangle = (N_1 - N_2)\mu_B. \tag{8.18}$$

We write $\mu_B B/k_B T = x$ and obtain

$$M = N\mu_B(e^x - e^{-x})/(e^x + e^{-x}) = N\mu_B \tanh(\mu_B B_0/k_B T). \tag{8.19}$$

In the low field (or high temperature) limit where $\mu_B B_0/k_B T \ll 1$, Equation (8.19) approximates to

$$M = N\mu_B^2 B_0/k_B T. \tag{8.20}$$

The susceptibility k, given by $k = M/B_0$, is then

$$k = N\mu_B^2/k_B T \tag{8.21}$$

which is constant, independent of the field. This constant, positive, field independent susceptibility is characteristic of a paramagnetic system. The susceptibility is inversely proportional to the absolute temperature, a relation known as Curie's Law. An example of Curie Law behaviour in a paramagnetic system is shown in Fig. 8.4. Measurement of the gradient of the plot of inverse susceptibility versus

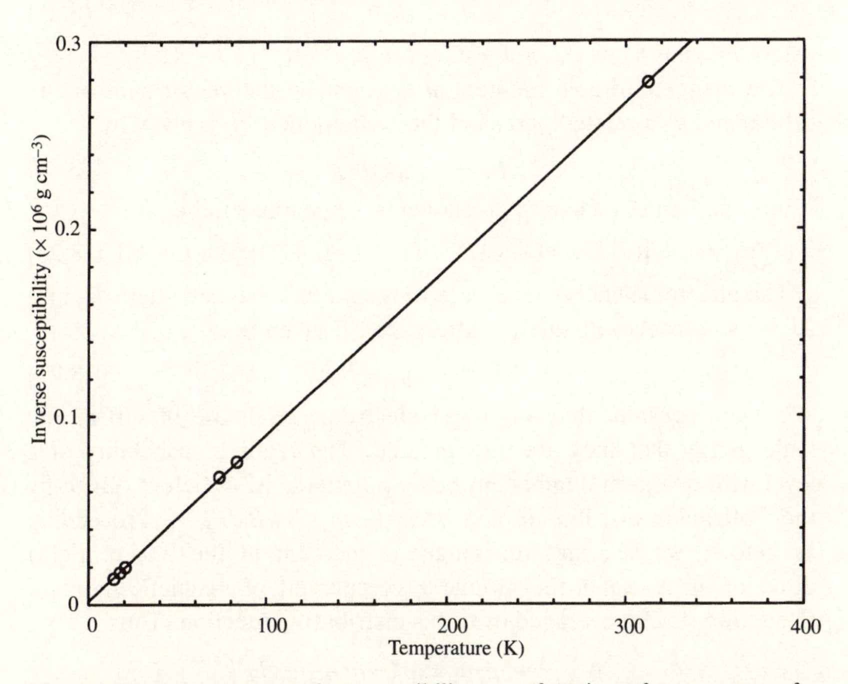

Fig. 8.4 Inverse paramagnetic susceptibility as a function of temperature for $CuO_4.KSO_4.6H_2O$ showing good agreement with Curie's Law. (After J. C. Hupse, *Physica* **9** (1942) 633.

temperature gives the magnetic moment per atom. In most cases, it is found not to correspond to $S = \frac{1}{2}$ as the orbital angular momentum of the electrons will contribute to the magnetic moment and hence the susceptibility.

8.3.2 Non-zero orbital angular momentum

In the general case the orbital angular momentum **L** and spin angular momentum **S** couple vectorially to give a total angular momentum **J** such that

$$\mathbf{J} = \mathbf{L} + \mathbf{S}. \tag{8.22}$$

As shown in any standard text on atomic physics, the absolute magnitude of the angular momentum is defined and the total angular momentum **J** has an expectation value given by

$$\mathbf{J}^2 = J(J+1)\hbar^2. \tag{8.23}$$

Similarly, the z component of angular momentum is defined and has an expectation value J_z given by

$$J_z = m_J\hbar \tag{8.24}$$

where m_J runs from $-J, -J+1, -J+2, \ldots$ to $\ldots J-1, J$.

The magnetic dipole moment $\boldsymbol{\mu}$ is given by the vector sum of the orbital and spin related parts and the z component μ_z is given by

$$\mu_z = -\mu_B g_J m_J \tag{8.25}$$

where the Landé g factor g_J is shown in Appendix 3 to be

$$g_J = 1 + [(J(J+1) + S(S+1) - L(L+1)]/[2J(J+1)]. \tag{8.26}$$

The electronic energy levels are thus split in a magnetic field B_0 into $2J+1$ sub-levels with energy difference δE given by

$$\delta E = m_J g_J \mu_B B_0. \tag{8.27}$$

We again consider that any other electronic levels are of sufficiently high energy that they are unpopulated. The relative probability of a level with quantum number m_J being populated is, as before, given by the Boltzmann distribution and is $\exp(-m_J g_J \mu_B B_0/k_B T)$. Proceeding as before, we see that the magnetic moment in the z (i.e. field) direction is N times the atomic z component of magnetic moment (Equation (8.25)) averaged over this distribution function. Thus

$$M = \frac{N\sum_{m_J} -m_J g_J \mu_B \exp(-m_J g_J \mu_B B_0/k_B T)}{\sum_{m_J} \exp(-m_J g_J \mu_B B_0/k_B T)}. \tag{8.28}$$

After much manipulation it can be shown that this is equal to

$$M = Ng_J J\mu_B B_J(g_J\mu_B B_0/k_B T) \tag{8.29}$$

where $B_J(x)$ is the Brillouin function given by

$$B_J(x) = [(2J + 1)/J] \coth \{[(2J + 1)/2J]x/2J\} - (1/2J) \coth (x/2J). \tag{8.30}$$

This expression gives an excellent fit to the measured magnetization at very low temperatures and very high fields (Fig. 8.5).

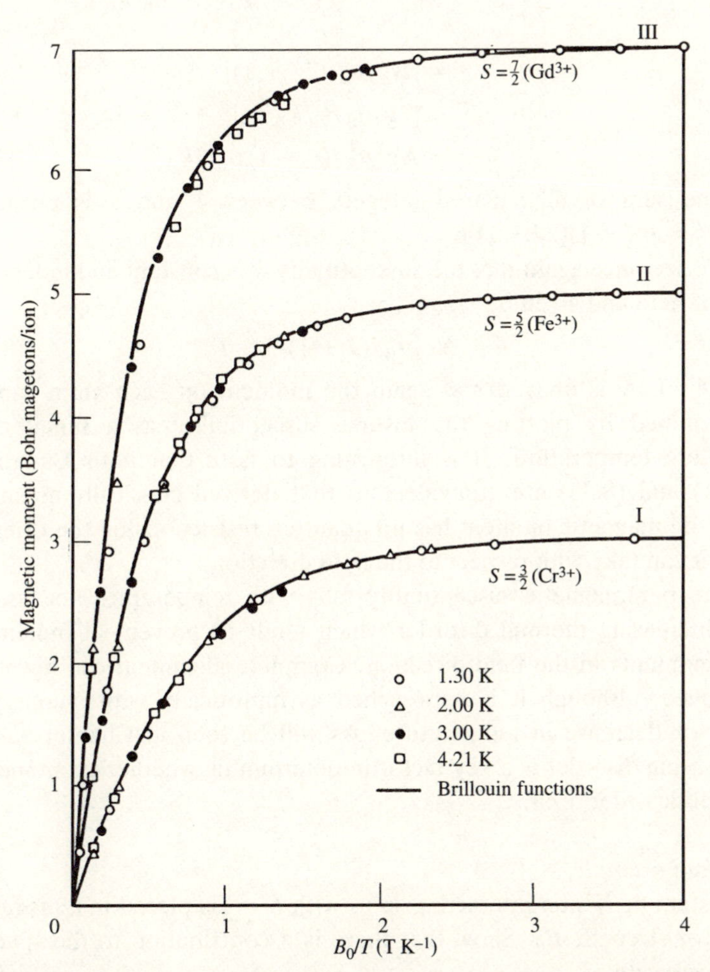

Fig. 8.5 Magnetization as a function of field for several paramagnetic salts at very low temperatures and very high fields. The continuous lines are fits to the Brillouin function. (I) Potassium chromium alum, (II) ferric ammonium alum and (III) gadolinium sulphate octahydrate. (After W. E. Henry, *Phys. Rev.* **88** (1952) 559.)

8.3.3 The low field limit

In low applied fields, and where the temperature is not extremely low, we may make the approximation that $m_J g_J \mu_B B_0/k_B T \ll 1$ and then

$$\exp(-m_J g_J \mu_B B_0/k_B T) = 1 - m_J g_J \mu_B B_0/k_B T. \quad (8.31)$$

Then

$$M = \frac{N\sum_{m_J} - m_J g_J \mu_B(1 - m_J g_J \mu_B B_0/k_B T)}{\sum_{m_J}(1 - m_J g_J \mu_B B_0/k_B T)}$$

$$= [N g_J \mu_B/(2J + 1)]$$
$$[2 g_J \mu_B B_0/k_B T(1^2 + 2^2 + 3^2 + \ldots + J^2)]$$
$$= N g_J^2 \mu_B^2 J(J + 1) B_0/3k_B T \quad (8.32)$$

as the sum of all squared integers between 1 and J is given by $\sum_1^J r^2 = J(J + 1)(2J + 1)/6$.

We see once again that the susceptibility k is constant and independent of field and given by

$$k = N g_J^2 \mu_B^2 J(J + 1)/3k_B T. \quad (8.33)$$

Curie's Law is obeyed and again the moment on each atom can be determined by plotting the inverse susceptibility as a function of absolute temperature. It is interesting to note that both Equations (8.21) and (8.33) are equivalent to that derived classically assuming that the magnetic moment has no quantum restrictions on the orientation it can take with respect to the field direction.

The paramagnetic susceptibility falls with temperature because of the increasing thermal disorder which tends to prevent alignment of the moments in the field direction. Complete alignment can never be achieved, although it is approached asymptotically with increase of field or decrease in temperature. As will be seen in Chapter 9, this increasing disorder is a key factor in determining whether spontaneous magnetic order occurs.

Worked example

A system of N non-interacting spins with $S = \frac{1}{2}$ is placed in a magnetic field of strength B_0. Show that there is a contribution to the specific heat capacity of

$$C_M = N k_B(2\mu_B B_0/k_B T)^2[\exp(2\mu_B B_0/k_B T)][1 + \exp(2\mu_B B_0/k_B T)]^{-2}. \quad (8.34)$$

Solution
Due to the magnetic energy term, the ground state energy level is split by $2\mu_B B_0$. The $m_s = -\frac{1}{2}$ state is lowered in energy by $-\mu_B B_0$ and the $m_s = +\frac{1}{2}$ state is raised by $+\mu_B B_0$. If we let N_1 and N_2 be the populations of the lower and upper levels respectively, we have, from Equations (8.17), that

$$N_1/N = [\exp(\mu_B B_0/k_B T)]/[\exp(\mu_B B_0/k_B T) + \exp(-\mu_B B_0/k_B T)] \tag{8.35a}$$

and

$$N_2/N = [\exp(-\mu_B B_0/k_B T)]/[\exp(\mu_B B_0/k_B T) + \exp(-\mu_B B_0/k_B T)]. \tag{8.35b}$$

The net energy associated with this magnetic splitting above that in the zero field situation is given by

$$E = (N_2 - N_1)\mu_B B_0. \tag{8.36}$$

The specific heat capacity associated with this magnetic energy is

$$C_M = \partial E/\partial T. \tag{8.37}$$

Thus,

$$\begin{aligned} C_M &= N\mu_B B_0 \partial[(e^{-x} - e^x)/(e^{-x} + e^x)]/\partial T \\ &= N\mu_B B_0(\partial x/\partial T)\partial[(e^{-x} - e^x)/(e^{-x} + e^x)]/\partial x \end{aligned} \tag{8.38}$$

where $x = \mu_B B_0/k_B T$. Explicitly,

$$\partial x/\partial T = -\mu_B B_0/k_B T^2 \tag{8.39}$$

and

$$\begin{aligned} &\partial[(e^{-x} - e^x)/(e^{-x} + e^x)]/\partial x \\ &\quad = \frac{-(e^{-x} + e^x)(e^{-x} + e^x) - (e^{-x} - e^x)(-e^{-x} + e^x)}{[e^{-x} + e^x]^2} \\ &\quad = \frac{-e^{-2x} - 2 - e^{2x} + e^{-2x} - 2 + e^{2x}}{(e^{-x} + e^x)^2} \\ &\quad = -4(e^{-x} + e^x)^{-2}. \end{aligned} \tag{8.40}$$

Therefore

$$\begin{aligned} C_M &= N\mu_B B_0(\mu_B B_0/k_B T^2)4[\exp(\mu_B B_0/k_B T) + \exp(-\mu_B B_0/k_B T)]^{-2} \\ &= Nk_B(2\mu_B B_0/k_B T)^2[\exp(2\mu_B B_0/k_B T)][1 + \exp(2\mu_B B_0/k_B T)]^{-2}. \end{aligned} \tag{8.41}$$

This is a specific case of a Schottky specific heat anomaly which has the form shown in Fig. 8.6. Where the splitting is too small for the peak to

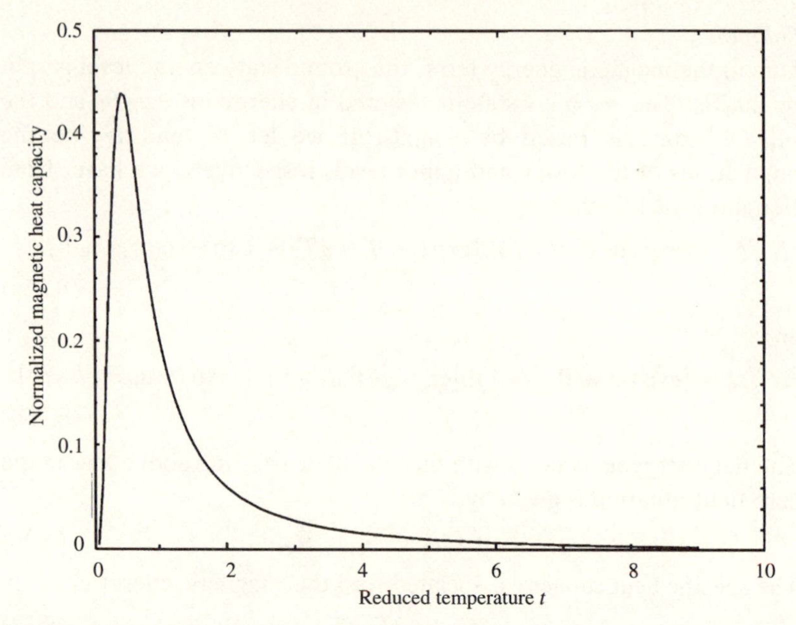

Fig. 8.6 Plot of Equation (8.41) showing the characteristic shape of the Schottky anomaly. Reduced temperature $t = k_B T/2\mu_B B_0$.

be observed at all but extremely low temperatures, it can be shown that the specific heat capacity has a characteristic T^{-2} tail. The Schottky anomaly is found wherever there are two or more closely spaced levels, the higher ones of which can have significant thermal equilibrium populations. Schottky anomalies are associated with non-interacting systems; where there are strong interactions, the specific heat capacity has an anomaly with shape like the Greek letter 'lambda' (λ). Such a lambda anomaly is characteristic of a cooperative phase transition (see Chapter 9).

8.4 Diamagnetism

Thus far we have considered electrons which are either acting as independent entities (as in the F centre) or are unpaired within the atomic structure. Such electrons have a net angular momentum and display paramagnetism. However, where there is a complete electronic shell, the net angular momentum is zero and thus closed shells do not contribute to the paramagnetic susceptibility. For example, there would be no effect from the noble gases such as argon or neon in either gaseous or solid state. However, all electrons do contribute to the magnetic moment in a small way and this phenomenon is known as diamagnetism.

The diamagnetic moment arises from the tendency of the electric charges to shield the interior of the body from the applied magnetic field. It is an example of Lenz's Law in that an induced current is set up to oppose the imposition of flux. In the microscopic case studied here, the currents arise from electronic motion. We can consider the situation from a classical viewpoint by noting that the motion of the electrons in the magnetic field is essentially the same as in zero field except for the Larmor precession. That is, the angular momentum vectors of each electron precess about the field direction with angular frequency

$$\omega = eB_0/2m. \tag{8.42}$$

This effect occurs whether or not the total angular momentum of the electron system in the atom is zero. The precessional motion is equivalent to an electric current I of magnitude

$$I = -(Ze/2\pi)(eB_0/2m) = -Ze^2B_0/4\pi m \tag{8.43}$$

where Z is the total number of electrons in the atom. The current loop is of area $\langle \rho^2 \rangle$ where

$$\langle \rho^2 \rangle = \langle x^2 \rangle + \langle y^2 \rangle. \tag{8.44}$$

This is the average of the square of the perpendicular distance of the electrons from the field (z) axis.

The associated magnetic dipole moment μ per atom is thus

$$\mu = -Ze^2B_0\langle \rho^2 \rangle/4m. \tag{8.45}$$

In terms of the mean square distance of the electrons from the nucleus $\langle r^2 \rangle$ defined by

$$\langle r^2 \rangle = \langle x^2 \rangle + \langle y^2 \rangle + \langle z^2 \rangle \tag{8.46}$$

and noting that by symmetry we have

$$\langle x^2 \rangle = \langle y^2 \rangle = \langle z^2 \rangle \tag{8.47}$$

we find that the magnetization M is given by

$$M = -NZe^2B_0\langle r^2 \rangle/6m \tag{8.48}$$

where N is the number of atoms per unit volume. The susceptibility k is thus

$$k = -NZe^2\langle r^2 \rangle/6m. \tag{8.49}$$

This is negative and independent of field and temperature. It is small compared with the paramagnetic susceptibility. When considered quantum mechanically and as a collective system, this result appears in addition to other components of the diamagnetic susceptibility. These are all small.

Problems

8.1 A crystal is formed of N molecules each of which has an unpaired electron in a triplet state such that $J = L = 1$. Assume no interactions between electrons of different molecules. By consideration of the relative equilibrium populations in each of the three levels formed under the application of a magnetic field, prove that the low field susceptibility k is given by

$$k = 2N\mu_B^2/3k_BT.$$

Compare this with that predicted by Equation (8.33).

8.2 A crystal is formed of N molecules each of which has two unpaired electrons in a triplet state such that $S = 1$ and $L = 0$. Again assume no interactions between electrons of different molecules. An $S = 0$ ground state exists at energy Δ below the triplet state, where $\Delta \ll k_BT$. Prove that the susceptibility k is given by

$$k = 2N\mu_B^2/k_BT.$$

Sketch the variation in energy levels as the applied field is increased. What will happen to the temperature of the solid if a very high field is removed adiabatically?

8.3 Consider the electron as a classical dipole of moment $\boldsymbol{\mu}$. Prove that, in a magnetic field B_0, the probability that the moment will lie within angular range θ to $\theta + d\theta$ of the field direction is proportional to $\exp(\mu B_0 \cos\theta/k_BT)2\pi \sin\theta d\theta$. Prove that the magnetization M is given by

$$M = N\mu[\coth(\mu B_0/k_BT) - k_BT/\mu B_0] = N\mu L(\mu B_0/k_BT)$$

where $L(x)$ is known as the Langevin function.

Use this result to prove that the low field susceptibility is given by

$$k = N\mu^2/3k_BT.$$

Compare this with the quantum mechanical results of Equations (8.21) and (8.33).

9

Magnetism

We have until now made use of the independent electron approximation, in which it is assumed that we can treat each electron independently of all of the others. In this chapter we will examine the consequence of the breakdown of this phenomenon.

It has been known for centuries, indeed it was known to the ancient Chinese, that magnetite or lodestone was attracted by the earth's field. Two pieces of lodestone attracted or repelled each other depending on which end of the lump of rock was pointed at the other. These chunks of material possess a spontaneous magnetic moment, i.e. they have a magnetization in zero external magnetic field.

We find that the elements iron, nickel and cobalt, bunched together in the middle of the periodic table, can also be induced to have a spontaneous moment at room temperature. The spontaneous magnetization **M** (defined as the magnetic moment per unit volume) is very large compared with that induced by a magnetic field in materials such as copper or zinc, which are very close in the periodic table. Alloys of iron, cobalt and nickel also have such properties which became known as ferromagnetism.

9.1 Basic phenomena

9.1.1 Hysteresis loops

Ferromagnetic materials show a characteristic $M-H$ (or $M-B_0$) loop. The susceptibility, defined by $k = M/B_0$ where B_0 is the external field, is very large and the magnetization displays hysteresis (Fig. 9.1). In sufficiently high field the magnetization saturates, this saturation magnetization being a characteristic of the material. The low field susceptibility, which is field dependent, is a function of the microstructure

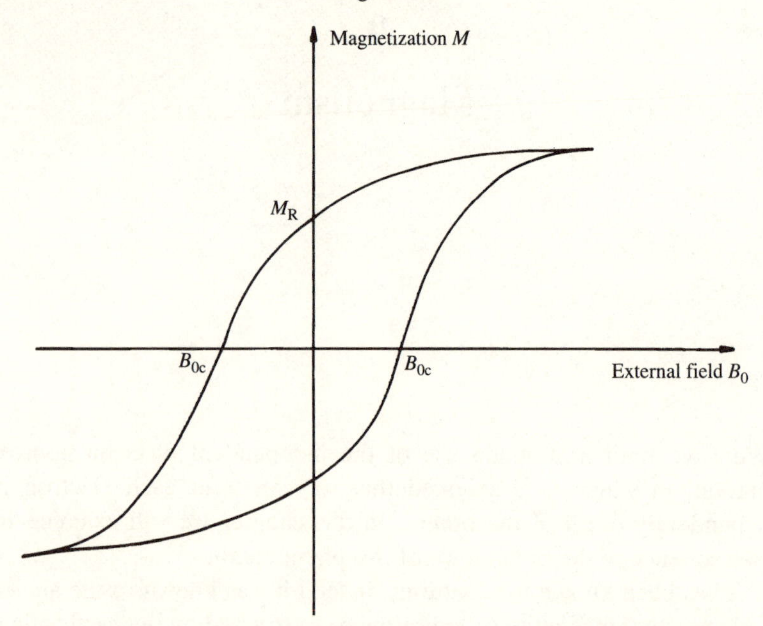

Fig. 9.1 Schematic $M-B_0$ hysteresis curve for a ferromagnetic material.

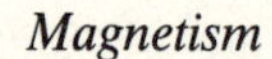

and history of the material. This low field k varies by several orders of magnitude. Note that the area of the magnetization loop (M versus B_0) is the work done per unit volume of material on cycling round the loop. Notice that there is, in general, a non-zero magnetization at zero applied field, the remnant magnetization M_R and that in a non-zero negative field, the coercive field or coercivity, B_{0c} is required in order to reduce the magnetization to zero. Cold work, i.e. plastic deformation, increases the hysteresis loop area and reduces the low field k. The origin of the hysteresis loop lies in the fact that in order to reduce the total free energy of the system, the material divides up into domains. These are regions where the direction of spontaneous magnetization differs. Thus although the material remains ordered at a microscopic level, the net measured magnetization (magnetic moment per unit volume) can be zero. Pinning of the domain walls is usually responsible for the origin of the hysteresis and the work dissipated on cycling around the $M-B_0$ loop. Domains form to minimize the total free energy of the system and the domain structure of a sample can be extremely complex and not necessarily reproducible. They are often of macroscopic dimensions, up to cubic millimetres in volume and in single crystals have orientation determined by the crystal structure.

9.1.2 Saturation magnetization

If we measure the saturation magnetization M_s as a function of temperature, all ferromagnetic materials show a similar characteristic behaviour; namely that the saturation magnetization falls as a function of temperature (Fig. 9.2). It extrapolates to zero at a characteristic temperature, known as the Curie temperature T_c, and above that temperature, the material behaves as a paramagnet. That means that the magnetization is small, positive and proportional to the external field.

9.1.3 Specific heat capacity

If we measure the specific heat capacity of a ferromagnet as it passes through the transition temperature, the Curie temperature, we find that there is a 'lambda' shaped anomaly in the specific heat (Fig. 9.3). The heat capacity C_p (at constant pressure) is the first differential of the entropy with respect to temperature. The additional heat capacity contribution C_m is

$$S_m = \int (C_m/T)\mathrm{d}T \tag{9.1}$$

where S_m is the magnetic entropy.

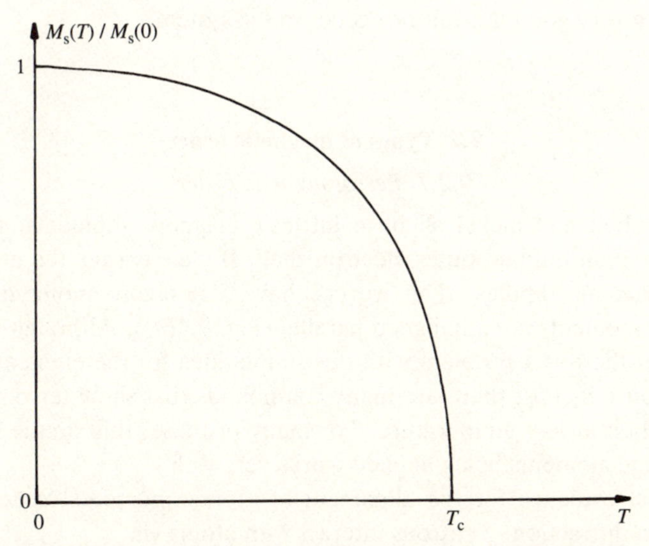

Fig. 9.2 Variation of the saturation magnetization M_s as a function of temperature T. Scaled to the value of M_s at $T = 0$.

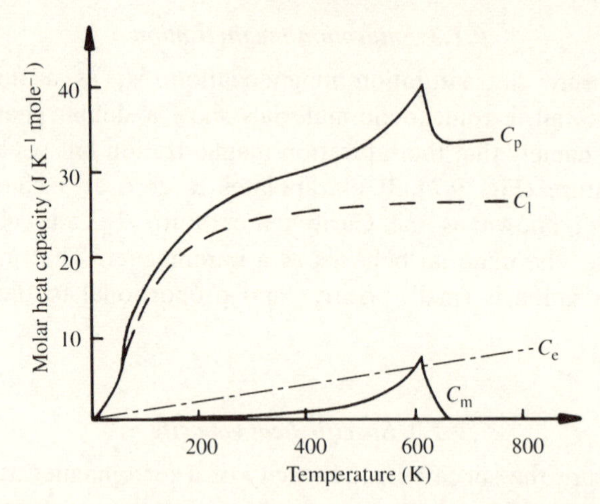

Fig. 9.3 Molar heat capacity C_p of nickel as a function of temperature showing lattice component C_l, electronic component C_e and magnetic component C_m.

The characteristic lambda shape to the curve indicates that the phase transition is a cooperative one. Unlike the Schottky anomaly resulting from the population of two states separated by a small energy difference which is rounded, the lambda anomaly gets sharper as the transition is approached. Above the transition, the approach is steeper. This is a very good indicator of order in the system.

9.2 Types of magnetic order

9.2.1 *Ferromagnetic order*

Iron, cobalt and nickel all have intrinsic magnetic moments; that is, they have an unfilled outer electron shell. If we envisage the atoms as tiny magnetic dipoles, then we can have a non-zero moment if the atomic moments are all aligned parallel (Fig. 9.4(*a*)). Although we will see that there is a problem with this simple idea for the elements, that is the pure metals, there are many compounds that show ferromagnetism, albeit at low temperature. For many of these, this simple picture of atomic moments being aligned works very well.

Direct evidence for the alignment of atomic moments comes from neutron diffraction. Neutrons interact with atoms via

1 the strong force, i.e. nuclear scattering

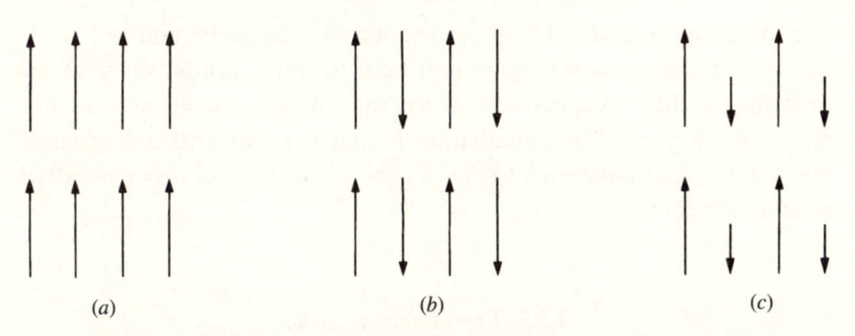

Fig. 9.4 Types of magnetic order. (*a*) Ferromagnetism, (*b*) antiferromagnetism, (*c*) ferrimagnetism.

2 the magnetic dipole moment of the neutron. This occurs only if the atom has a magnetic moment and therefore there is a magnetic dipole–dipole interaction.

Both interactions contribute to the total scattering cross-section. If the atoms are arranged in a crystal lattice, the nuclear scattering is coherent and at certain angles, where Bragg's Law is satisfied, we observe very strong scattered intensity. If the sample is a powder, and we place the sample in a beam of monochromatic neutrons, then we observe scattering into cones at scattering angles 2θ corresponding to

$$\lambda = 2d_{hkl}\sin\theta \tag{9.2}$$

where d_{hkl} are the atomic plane spacings. The magnetic dipole–dipole interaction is a vector interaction and depends on the relative orientations of the neutron and atomic moments. If the atomic moments are oriented randomly, the scattering is incoherent; but if the atomic moments are aligned, then *magnetic Bragg scattering* is observed. If we measure a diffraction pattern as the crystal cools through the Curie temperature, we get a sharp increase in the diffracted intensity as the magnetization rises rapidly below the Curie temperature. The increase in intensity has the same characteristic shape as the saturation magnetization with temperature.

9.2.2 Antiferromagnetic order

In materials such as NiO or $KNiF_3$ we find experimentally that neutrons often show order which is not ferromagnetic and with which there is no associated spontaneous magnetization measured macroscopically. What is found is that extra Bragg peaks appear below the transition temperature. These correspond to a magnetic unit cell *twice* the size of the crystallographic unit cell. In other words, we have an ordering of the moments with twice the atomic spacing and no net magnetic moment. The explanation is that the moments are ordered not parallel, but *antiparallel* (Fig. 9.4(*b*)). This type of order is called antiferromagnetism.

9.2.3 Ferrimagnetic order

For some materials neutrons show both extra peaks and increased intensity in the normal Bragg peaks, indicating the presence of a spontaneous moment. These materials also show a spontaneous magnetic moment in the bulk state and for practical purposes can be regarded as equivalent to ferromagnets. This phenomenon is known as ferrimagnetic order and it corresponds to two antiparallel sublattices of different sublattice moment (Fig. 9.4(*c*)). Clearly, the net moment is non-zero.

9.3 The exchange interaction

The key question which we must address is how these moments can align spontaneously. A first suggestion might be that the dipole–dipole interaction between the moments is large enough to overcome the effect of thermal disorder and thus lead to ordering. Unfortunately, the energy associated with the magnetic ordering has to be about $k_B T_c$ and the magnetic interaction between the dipoles is simply not sufficiently strong. Although Weiss developed a very successful phenomenological theory based on an internal 'molecular field' of unknown origin at the beginning of the century, it was not until the advent of quantum mechanics that the origin of the interaction became clear. The interaction responsible for magnetic order is the *exchange interaction*. This interaction results from the fact that electrons are identical and obey the Pauli exclusion principle; that is the overall wavefunction must be antisymmetric. Note that the exchange interaction is electrostatic.

Let us begin by considering the hydrogen molecule. We have two electrons moving in similar potential fields. If the interaction between electrons is neglected, then the Schrödinger equation is

$$\left[-\frac{\hbar^2}{2m^*}(\nabla_1^2 + \nabla_2^2) + V(q_1) + V(q_2)\right]\psi = E\psi, \tag{9.3}$$

where q_1 and q_2 are generalized coordinates of electrons 1 and 2. Possible solutions are $\psi_a(1)\psi_b(2)$ or $\psi_b(1)\psi_a(2)$ with

$$E = E_a + E_b. \tag{9.4}$$

The term $\psi_a(1)$ is the single electron wavefunction when electron 1 is in state a and it is a solution of the one electron Schrödinger equation; $\psi_b(2)$ is the solution for electron 2 in state b. Similarly, $\psi_b(1)$ is the wavefunction for electron 1 in state b and $\psi_a(2)$ for electron 2 in state a. As the electrons are indistinguishable, the total wavefunction $\psi(1, 2)$ must be such that

$$|\psi(1, 2)|^2 \mathrm{d}q_1 \mathrm{d}q_2 = |\psi(2, 1)|^2 \mathrm{d}q_1 \mathrm{d}q_2 \tag{9.5}$$

and thus either

$$\psi(1, 2) = +\psi(2, 1) \tag{9.6a}$$

or

$$\psi(1, 2) = -\psi(2, 1). \tag{9.6b}$$

The first total wavefunction is symmetric, the second is antisymmetric. Linear combinations of the single particle wavefunctions satisfying these requirements are

$$\psi_{\mathrm{sym}}(1, 2) = [\psi_a(1)\psi_b(2) + \psi_b(1)\psi_a(2)]/2^{1/2} \tag{9.7a}$$

$$\psi_{\mathrm{anti}}(1, 2) = [\psi_a(1)\psi_b(2) - \psi_b(1)\psi_a(2)]/2^{1/2}. \tag{9.7b}$$

We can separate out the spin and orbital components of the wavefunction if the orbital moment is quenched and only the spin contributes to the moment. (Quenching arises because the electrons in the 3d transition metal series are the outer electrons and move in a potential which has the symmetry of the crystal lattice, not a central potential. Orbital angular moment then does not remain a good quantum number and the orbital component of the magnetic moment averages to zero.) This separation of variables gives

$$\psi = \phi(r)\chi \tag{9.8}$$

where $\phi(r)$ is the solution for an electron without spin and χ is the spin component. Remembering that the overall wavefunction must be antisymmetric, we can write the total wavefunction as either

$$\psi(1, 2) = \phi_{\mathrm{sym}}(1, 2)\chi_{\mathrm{anti}}(1, 2) \tag{9.9a}$$

or

$$\psi(1, 2) = \phi_{\mathrm{anti}}(1, 2)\chi_{\mathrm{sym}}(1, 2). \tag{9.9b}$$

Let us now consider the interaction between the electrons. Take the Hamiltonian H as

$$H = e^2[(1/r_{ab}) + (1/r_{12}) - (1/r_{a2}) - (1/r_{b1})] \tag{9.10}$$

where r_{ab} is the distance between nuclei, r_{12} is the distance between electrons and r_{a2} and r_{b1} are the distances between electrons and nuclei of the other atom. Recalling that the energy

$$E = \int \psi^* H \psi \mathrm{d}\tau \tag{9.11}$$

we can determine the additional energy of the two electron system from the non-interacting case. Explicitly we have, remembering that H does not act on the spin part of the wavefunction,

$$E_1 = A^2(K_{12} + J_{12}) \quad \text{singlet state } S = 0 \tag{9.12a}$$

$$E_2 = B^2(K_{12} - J_{12}) \quad \text{triplet state } S = 1 \tag{9.12b}$$

where A and B are normalizing factors and

$$K_{12} = \iint \phi_a^*(1)\phi_b^*(2) H_{12} \phi_b(2)\phi_a(1) \mathrm{d}\tau_1 \mathrm{d}\tau_2 \tag{9.13a}$$

$$J_{12} = \iint \phi_a^*(1)\phi_b^*(2) H_{12} \phi_a(2)\phi_b(1) \mathrm{d}\tau_1 \mathrm{d}\tau_2. \tag{9.13b}$$

The term J_{12} is called the exchange integral. For the hydrogen molecule J is negative and hence the ground state is antiparallel. We might thus expect all magnetically ordered states to be antiferromagnetic and indeed this is the case for many of them. However, if the wavefunctions ϕ_a and ϕ_b have no nodes in the region of appreciable overlap, then $\phi_a^*(1)\phi_b^*(2)\phi_a(2)\phi_b(1)$ is positive. The sign of the exchange integral will be positive if the positive terms in the Hamiltonian exceed the negative ones. In our Hamiltonian, a large value of e^2/r_{12} occurs if the wavefunctions are large only mid-way between the nuclei, since only in this region is r_{12} small. Further, the terms $-e^2/r_{a2}$ and $-e^2/r_{b1}$ are smallest for wavefunctions that are small near the nuclei. Thus if the interatomic spacing r_{ab} is large compared with the radii of the orbitals, that is the ionic radii of the atoms, then J can be positive and ferromagnetic order results. We need now to generalize this result to a system of many spins. To do this we note that the operator $-\frac{1}{2} - \mathbf{s}_1 \cdot \mathbf{s}_2$ is $+1$ and -1 for the triplet and singlet states respectively. (This can be seen by noting that $\mathbf{S} = \mathbf{s}_1 + \mathbf{s}_2$ and that $\mathbf{S}^2$, $\mathbf{s}_1^2$ and $\mathbf{s}_2^2$ are eigenfunctions and working out the eigenvalues.) Thus we can write

$$E = K - \tfrac{1}{2}J - 2J_{\mathrm{e}}\mathbf{s}_1 \cdot \mathbf{s}_2, \tag{9.14}$$

where J_{e} is the usual nomenclature for the exchange integral. The first two terms are independent of the spin orientation and we usually use what is called the *Heisenberg Hamiltonian*

$$H = -2J_e \sum_{\text{all neighbours}} \mathbf{s}_i \cdot \mathbf{s}_j \tag{9.15}$$

to describe the energy of the system.

The important thing to note about this formula is that, although it arises from an electrostatic interaction, it has the form of the magnetic dipole–dipole interaction, namely $\mathbf{s}_i \cdot \mathbf{s}_j$. Thus, we can consider the interaction *as if* it were the interaction of a spin in a magnetic field created by the other spins. The exchange interaction is found to be many orders of magnitude larger than the magnetic dipole–dipole interaction and large enough to account for the value of the Curie temperatures. We will see, however, that the simple direct exchange interaction considered above is not always applicable. Let us, nevertheless, use the result to look at how magnetic solids order.

9.4 Mean field theory of ferromagnetism

This is a very powerful theory and the modern equivalent of the 'molecular field' theory developed by Weiss in 1907. It assumes that the exchange interaction results in an effective magnetic field at an ion and that this field is the mean field produced by all the other spins. In other words, we can consider one spin in the mean field of all the others. This mean field will be proportional to the magnetization. Let us write this interaction field $\mathbf{B}_E$ as

$$\mathbf{B}_E = \lambda \mathbf{M} \tag{9.16}$$

where λ is a constant, independent of temperature.

In the paramagnetic, disordered state, an applied field $\mathbf{B}_0$ leads to a magnetization $\mathbf{M}$. If k_p is the intrinsic paramagnetic susceptibility we have

$$M = k_p(B_0 + B_E). \tag{9.17}$$

We know, from Equation (8.33) that

$$k_p = C/T \tag{9.18}$$

where C is the Curie constant given by

$$C = N g_J^2 J(J+1)\mu_B^2 / 3k_B$$

and where g_J is the Landé g factor. The effective moment $p_{(\text{eff})}$ is given by

$$p_{(\text{eff})} = g_J \mu_B [J(J+1)]^{1/2}. \tag{9.19}$$

Thus we have

$$MT = C(B_0 + \lambda M) \tag{9.20}$$

and the measured susceptibility k is given by

$$k = M/B_0 = C/(T - \lambda C). \tag{9.21}$$

This equation has a singularity at $T = \lambda C = T_c$; *if k is infinite there exists a spontaneous magnetization in zero field*. It is usual to plot $1/k$ versus T to determine the effective moment and we note that from Equation (9.21), we expect $1/k$ to become zero at T_c (Fig. 9.5). From T_c we can determine the value of the mean field B_E which we find for iron to be about 1000 tesla!

9.4.1 Temperature dependence of saturation magnetization

In the above derivation, we assumed the high temperature approximation for the magnetization as a function of field, namely the Curie Law. The proper expression for the magnetization M of an assembly of moments J is

$$M = Ng_J\mu_B JB_J(x) \tag{9.22}$$

where $x = Jg_J\mu_B(B_0 + B_E)/k_BT$ and $B_J(x)$ is the Brillouin function

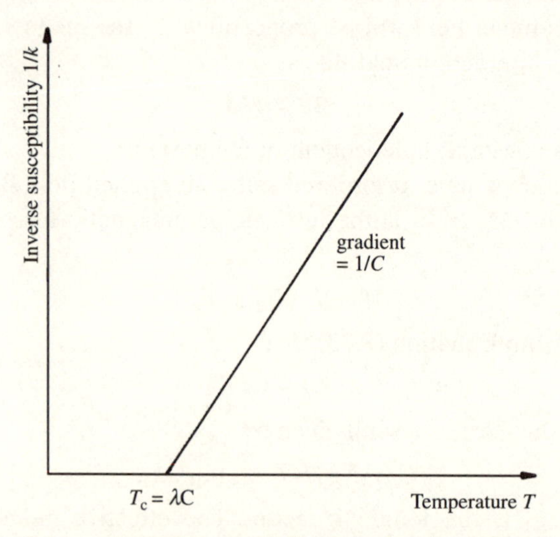

Fig. 9.5 Plot of the reciprocal susceptibility as a function of temperature for a material exhibiting ferromagnetic interactions and ferromagnetic order below $T_c = \lambda C$.

given by

$$B_J(x) = [(2J+1)/J] \coth \{[(2J+1)/2J]x/2J\} - (1/2J) \coth (x/2J). \tag{9.23}$$

Now in zero external field in the ordered state, the field B is λM, the mean interaction field. Thus we have

$$M = N g_J \mu_B J B_J(x) \tag{9.24}$$

where $x = J g_J \mu_B \lambda M / k_B T$.

We can solve this equation graphically by looking for the intersection of the curves given by the left and right hand sides of this equation. Let us illustrate this by setting $J = S = \frac{1}{2}$. Then

$$M = N \mu_B \tanh (\mu_B \lambda M / k_B T). \tag{9.25}$$

Writing the reduced magnetization $m = M(T)/M(0)$ and the reduced temperature $t = T/T_c$ we have

$$m = \tanh (m/t). \tag{9.26}$$

We plot both left and right hand sides against m (Fig. 9.6(*a*)). The left hand side is a line of unity gradient. Note that for $t = 1$, this line is tangential to the right hand side line. Generally the intersection gives the value of the spontaneous (saturation) magnetization as a function of temperature. This we can plot as a function of reduced temperature t and obtain curves which have a very good general agreement with those observed experimentally (Fig. 9.6(*b*)). Note that there is very little difference between curves for $J = \frac{1}{2}$ and $J = 1$ but that the curve for large J is significantly different.

Behaviour of M close to T_c

Here x is small and the Brillouin function can be expanded to give

$$B_J(x) = [(J+1)/3J]x - [(J+1)/3J][(2J^2+2J+1)/30J^2]x^3. \tag{9.27}$$

Substitution of $x = J g_J \mu_B \lambda M / k_B T$ yields, for $T \approx T_c$

$$[M(T)/M(0)]^2 = 10(J+1)^2(1 - T/T_c)/3[J^2 + (J+1)^2]. \tag{9.28}$$

The experimental results are in quite good agreement near T_c, there being about 1% difference between theory and experiment.

Behaviour of M close to $T = 0$

If we let t tend to zero, then t becomes very small and x is large. Noting that $\tanh x = 1 - 2 \exp(-2x)$ for large x, we have

$$m(t) = 1 - \{\exp[-3t/(J+1)]/J\} \tag{9.29}$$

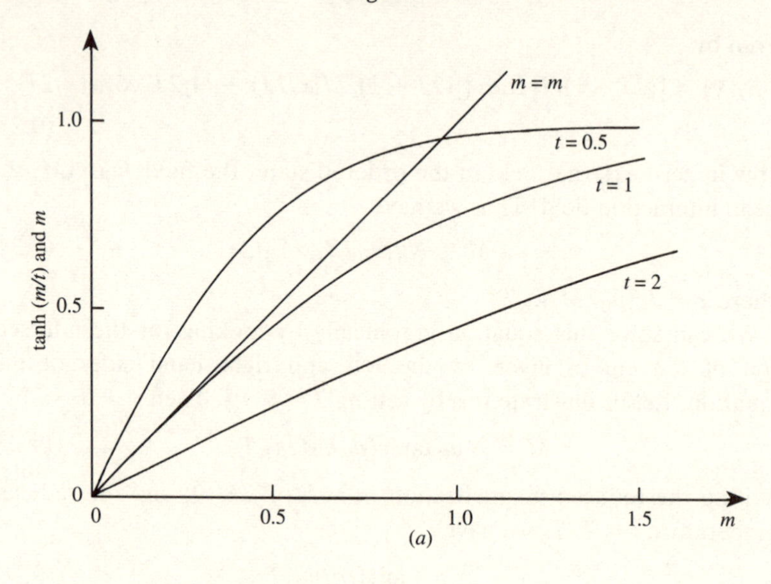

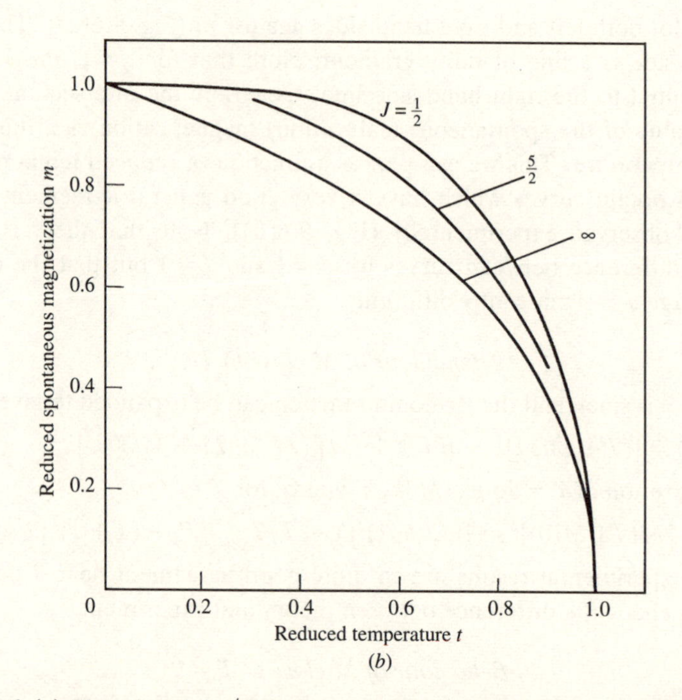

Fig. 9.6 (a) Plot of $\tanh(m/t)$ and m versus m to demonstrate the graphical solution of Equation (9.26). Above a reduced temperature $t = 1$, no intersection occurs and spontaneous order does not exist. (b) Plot of m versus t for various values of J.

as t tends to 0. However, we find experimentally that the dependence close to absolute zero varies as

$$m = 1 - AT^{3/2} \tag{9.30}$$

where A is a constant. This discrepancy is a strong argument for collective excitations called *spin waves* which we will meet later in the chapter. The low temperature region is one where the mean field theory does not hold well.

9.5 Mean field theory of antiferromagnetism

The same approach as that above can be taken for antiferromagnets where there are two sublattices with equal magnitude spins on equivalent sites. Consider a BCC lattice with a atoms on the corner sites and b atoms on the body positions. Atom a has all b sites as nearest neighbours and vice versa. The mean interaction field for spin a is now written

$$B_a = -\lambda_{aa}M_a - \lambda_{ab}M_b \tag{9.31}$$

where M_a and M_b are the magnetizations of the two sublattices of a and b spins. A mean field constant λ_{aa} describes the interaction strength between spins of type a and a mean field constant λ_{ab} describes the interaction between a and b spin systems. Note that the mean field in both cases is antiparallel to the applied field, as implied by the minus signs.

As the lattice sites are equivalent, we can write for the b spins

$$B_b = -\lambda_{ba}M_a - \lambda_{bb}M_b \tag{9.32}$$

where $\lambda_{ab} = \lambda_{ba}$ and $\lambda_{aa} = \lambda_{bb} = \lambda_{ii}$. Thus the total fields seen by atoms of type a and b in the mean field approximation are

$$B_a = B_0 - \lambda_{ab}M_b - \lambda_{ii}M_a \tag{9.33a}$$

$$B_b = B_0 - \lambda_{ii}M_b - \lambda_{ab}M_a. \tag{9.33b}$$

Note that λ_{ii} can be either positive or negative, but that for antiferromagnetic order, λ_{ab} must be positive as defined here.

In thermal equilibrium the magnetization of sublattice a is

$$M_a = Ng_S\mu_B SB_S(x_a)/2 \tag{9.34}$$

where $B_S(x_a)$ is the Brillouin function for spin S and $x_a = Sg_S\mu_B B_a/k_B T$ and similarly for the magnetization of sublattice b.

9.5.1 Behaviour in the paramagnetic region

Here, as with the ferromagnetic case, we can approximate the Brillouin function to $g_S\mu_B(S+1)/3k_BT$ so that

$$M_a = Ng_S^2\mu_B^2S(S+1)B_a/6k_BT \tag{9.35a}$$

and

$$M_b = Ng_S^2\mu_B^2S(S+1)B_b/6k_BT. \tag{9.35b}$$

Substitution for B_a and B_b gives

$$M_a = Ng_S^2\mu_B^2S(S+1)(B_0 - \lambda_{ab}M_b - \lambda_{ii}M_a)/6k_BT \tag{9.36a}$$

and

$$M_b = Ng_S^2\mu_B^2S(S+1)(B_0 - \lambda_{ab}M_a - \lambda_{ii}M_b)/6k_BT. \tag{9.36b}$$

Thus as $M = M_a + M_b$ we have

$$M = Ng_S^2\mu_B^2S(S+1)[2B_0 - (\lambda_{ii} + \lambda_{ab})M]/6k_BT. \tag{9.37}$$

Hence the susceptibility $k = M/B_0$ is given by

$$k = 2C/(T + \Theta) \tag{9.38}$$

where $C = Ng_S^2\mu_B^2S(S+1)/6k_B$ and $\Theta = C(\lambda_{ab} + \lambda_{ii})$.

Note that Θ is positive and if we plot k^{-1} versus T (Fig. 9.7) we get a plot very similar to that of the Curie–Weiss equation (Equation (9.21)) except that the intercept with the T axis is negative!

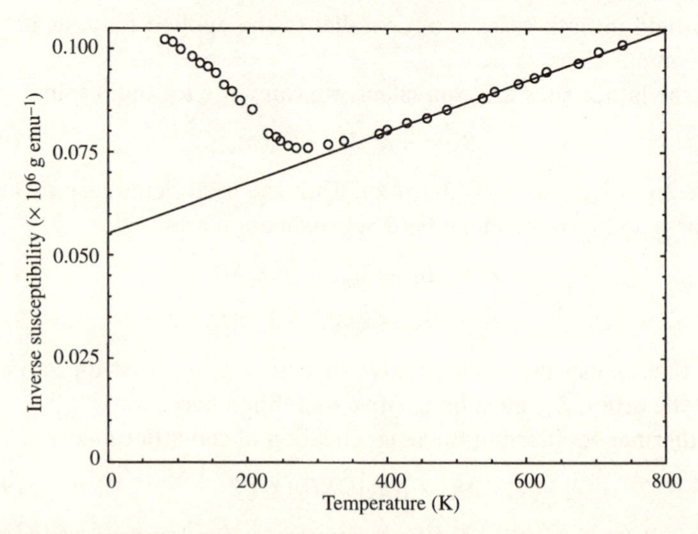

Fig. 9.7 Curie–Weiss plot of the reciprocal susceptibility as a function of temperature for $KNiF_3$ which has an antiferromagnetic transition at 246 K

9.5.2 The Néel temperature

Clearly the straight line cannot go on below zero and thus ordering cannot occur at the intercept as in the ferromagnetic case. We find, however, that there is a temperature T_N called the Néel temperature below which the sublattices spontaneously order. Below the Néel temperature, M_a and M_b are non-zero in zero applied field. We have already

$$M_a = C(-\lambda_{ab}M_b - \lambda_{ii}M_a)/T_N \tag{9.39a}$$

$$M_b = C(-\lambda_{ii}M_b - \lambda_{ab}M_a)/T_N \tag{9.39b}$$

and for there to be a non-trivial solution, the determinant of the coefficients must vanish. Thus

$$\begin{vmatrix} 1 + \lambda_{ii}C/T_N & C\lambda_{ab}/T_N \\ C\lambda_{ab}/T_N & 1 + \lambda_{ii}C/T_N \end{vmatrix} = 0 \tag{9.40}$$

and we find that this occurs at

$$T_N = C(\lambda_{ab} - \lambda_{ii}). \tag{9.41}$$

The Néel temperature is higher the stronger the *ab* interaction and the weaker the *aa* or *bb* interaction. It follows that

$$\Theta/T_N = (\lambda_{ab} + \lambda_{ii})/(\lambda_{ab} - \lambda_{ii}). \tag{9.42}$$

If $\lambda_{ii} = 0$, then $T_N = \Theta$.

9.5.3 Susceptibility below the Néel temperature

Parallel susceptibility

In zero applied field, the sublattice magnetizations are antiparallel. Anisotropy will be ignored for the moment except in so far as it defines a z direction. Suppose we now have a field applied parallel to the z axis, i.e. parallel and antiparallel to the sublattice magnetizations. We recall that

$$M_a = Ng_S\mu_B S B_S(x_a)/2 \tag{9.43}$$

with $x_a = Sg_S\mu_B B_a/k_B T$ and similarly for M_b. Writing, for $B_0 = 0$, $M_a = -M_b = M_0$, we have

$$x_0 = g_S\mu_B S(\lambda_{ab} - \lambda_{ii})M_0/k_B T. \tag{9.44}$$

We can expand $B_S(x)$ as a Taylor series in B_0 and then to first order

$$B_S(x_a) = B_S(x_0) + g_S\mu_B S \times [B_0 + \lambda_{ii}(M_0 - M_a) + \lambda_{ab}(M_b - M_0)]B_S'(x_0)/k_B T \tag{9.45a}$$

$$B_S(x_b) = B_S(x_0) - g_S\mu_B S \times [B_0 + \lambda_{ab}(M_0 - M_a) + \lambda_{ii}(M_b - M_0)]B'_S(x_0)/k_B T \quad (9.45b)$$

where $B'_S(x_0)$ is the derivative of the Brillouin function with respect to x. Multiplication by $g_S\mu_B SN/2$ yields the sublattice magnetizations. The induced magnetization $M = M_a - M_b$ is the difference in the two equations and the parallel susceptibility $k_\parallel$ is therefore

$$k_\parallel = N\mu_B^2 g_S^2 S^2 B'_S(x_0)/[k_B T + (\lambda_{ii} + \lambda_{ab})\mu_B^2 g_S^2 S^2 N B'_S(x_0)/2]. \quad (9.46)$$

At absolute zero $k_\parallel$ is zero while, for $T = T_N$ the susceptibility must equal that in the paramagnetic phase. Thus $k_\parallel$ as a function of T has the form shown in Fig. 9.8.

Perpendicular susceptibility

If the field B is applied perpendicular to the sublattice magnetization, a torque exists which tends to rotate the sublattice magnetizations. The rotation is opposed by the mean interaction field. In equilibrium

$$\mathbf{M}_a \times (\mathbf{B}_0 - \lambda_{ab}\mathbf{M}_b - \lambda_{aa}\mathbf{M}_a) = 0$$

or

$$\mathbf{M}_a \times (\mathbf{B}_0 - \lambda_{ab}\mathbf{M}_b) = 0. \quad (9.47)$$

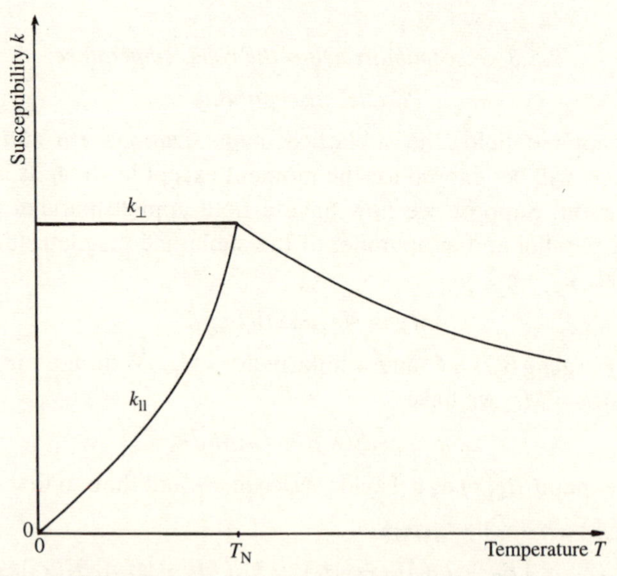

Fig. 9.8 Susceptibilities parallel ($k_\parallel$) and perpendicular ($k_\perp$) to the ordering direction of a uniaxial antiferromagnet below the Néel temperature.

Hence, if ϕ is the angle between the applied field and the sublattice magnetization direction,

$$M_a B_0 \cos\phi = \lambda_{ab} M_a M_b \sin 2\phi, \tag{9.48}$$

which becomes

$$2M_b \sin\phi = B_0/\lambda_{ab}. \tag{9.49}$$

The net magnetization in the field direction is $M = (M_a + M_b)\sin\phi$ and as $M_a = M_b$ we have for the perpendicular susceptibility $k_\perp$

$$k_\perp = 1/\lambda_{ab}. \tag{9.50}$$

This is independent of temperature as sketched in Fig. 9.8 and is equal to the value of the parallel susceptibility at the Néel temperature. In a polycrystalline sample there will be a mixture of orientations and the measured susceptibility will be an average of the two values above. There are two orthogonal axes perpendicular to the z direction and therefore we add the weighting of 2:1 for the perpendicular and parallel components. Hence

$$\langle k \rangle = k_\parallel/3 + 2k_\perp/3. \tag{9.51}$$

9.6 Mean field theory of ferrimagnetism

The requirement that the sublattice magnetizations should be of equal magnitude is a restriction which is not totally necessary. Suppose that the a and b sites are not equivalent and that $M_a \neq M_b$. This implies that the mean field constants λ_{aa} and λ_{bb} are now different, although by symmetry $\lambda_{ab} = \lambda_{ba}$. The total mean fields are then

$$B_a = B_0 - \lambda_{aa} M_a - \lambda_{ab} M_b \tag{9.52a}$$

and

$$B_b = B_0 - \lambda_{ab} M_a - \lambda_{bb} M_b. \tag{9.52b}$$

We can proceed as before to evaluate the susceptibility in the paramagnetic region. With Curie constants C_a and C_b for the a and b sublattices respectively we have $M_a = C_a B_a/T$ and $M_b = C_b B_b/T$ and it follows that

$$(T + C_a\lambda_{aa})M_a + C_a\lambda_{ab}M_b = C_a B_0 \tag{9.53a}$$

and

$$C_b\lambda_{ab}M_a + (T + C_b\lambda_{bb})M_b = C_b B_0. \tag{9.53b}$$

Solving and noting again that $M = M_a + M_b$, we find that the susceptibility is of the form

$$1/k = (T/C) - (1/k_0) - [\sigma/(\mathrm{T} - \Theta')] \tag{9.54}$$

where

$$\begin{aligned} C &= C_a + C_b \\ 1/k_0 &= -(C_a^2\lambda_{aa} + C_b^2\lambda_{bb} + 2C_aC_b\lambda_{ab})/C^2 \\ \Theta' &= -C_aC_b(\lambda_{aa} + \lambda_{bb} - 2\lambda_{ab})/C \\ \sigma &= C_aC_b\{C_a^2(\lambda_{aa} - \lambda_{ab})^2 + C_b^2(\lambda_{bb} - \lambda_{ab})^2 \\ &\quad - 2C_aC_b[\lambda_{ab}^2 - (\lambda_{aa} + \lambda_{bb})\lambda_{ab} + \lambda_{aa}\lambda_{bb}]\}/C^3. \end{aligned}$$

This messy equation has the form

$$k = C/(T + \Theta) \tag{9.55}$$

for large T. As usual this is plotted in the form of $1/k$ versus T and for high T we get a straight line (Fig. 9.9). There is a negative T axis intercept at Θ which is given by $\Theta = -C/k_0$.

9.6.1 Ferrimagnetic Néel temperature

As with the antiferromagnet we find that there is a temperature at which non-zero values of M_a and M_b can occur for zero B_0. Let $B_0 = 0$ and set the determinant of the coefficients to zero and we find $T_{\rm NF}$ given by

$$T_{\rm NF} = -(C_a\lambda_{aa} + C_b\lambda_{bb})/2 + [(C_a\lambda_{aa} - C_b\lambda_{bb})^2 + 4C_aC_b\lambda_{ab}{}^2]^{1/2}/2. \tag{9.56}$$

The curvature on the $1/k$ versus T plot in the vicinity of the Néel temperature, seen very clearly in Fig. 9.9, is a characteristic of a ferrimagnet.

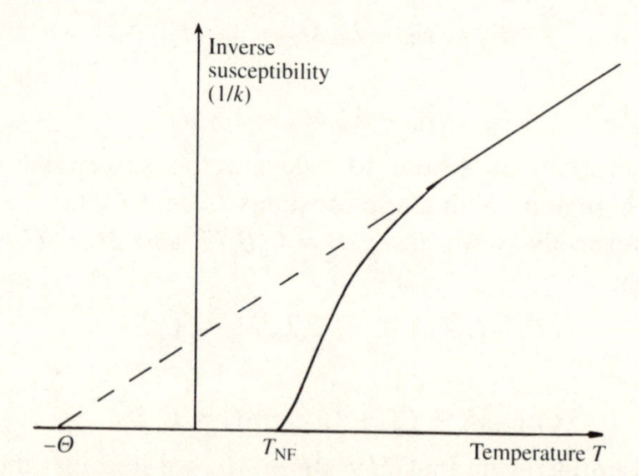

Fig. 9.9 Plot of reciprocal susceptibility as a function of temperature for a ferrimagnet. Note the negative intercept of the linear region and downward curvature to the ferrimagnetic Néel temperature.

9.6.2 Magnetization below the Néel temperature

Calculation of the spontaneous magnetization below T_{NF} is straightforward but tiresome. The procedure is to solve coupled equations for the two sublattices under the constraint that $M = M_a + M_b$. It is best to solve the equations using a digital computer. There are some curious results, depending on the relative magnitudes of the mean field parameters. Examples of M versus T curves for ferrimagnets for various values of λ_{ab}, λ_{aa} and λ_{bb} are shown in Fig. 9.10. Probably the most important ferrimagnets are garnets. They have the general formula $\{P_3\}(Q_2)R_3O_{12}$. The basic crystal structure is cubic with 96 oxygen

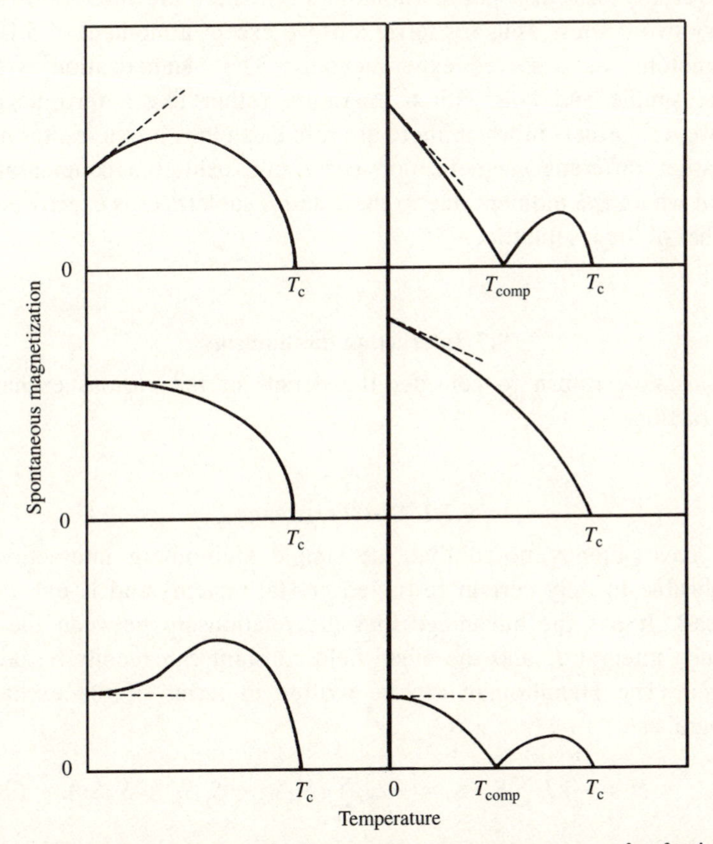

Fig. 9.10 Spontaneous magnetization versus temperature curves for ferrimagnets having various values of the mean field constants λ_{ab}, λ_{aa} and λ_{bb}. All forms of M versus T curve have now been found experimentally in two sublattice systems. T_{comp} corresponds to the temperature where the magnetization of the two sublattices is equal and opposite. (After L. Néel, *Ann. Phys.* (1948) 154.)

atoms per unit cell and they are important in microwave devices and bubble memories as well as being gemstones. There are three sublattices and mean field calculations can be performed in the manner shown above, but with six mean field parameters. The paramagnetic susceptibility has the form

$$1/k = (T/C) - (1/k_0) - (\sigma T + D)/(T^2 - \Theta' T + f), \quad (9.57)$$

with D and f constants. In garnets such as YIG (yttrium iron garnet), there are moments only due to the iron atoms which are trivalent. There are two types of sites; sublattices a and c are associated with octahedral sites while sublattice d is associated with tetrahedral sites. Iron atoms exist in a and d sublattices but there are three d sites to every two a sites. Thus for ferric ions we except a moment of 5 Bohr magnetons, as observed experimentally. The magnetization is also quite simple and falls with temperature rather like a ferromagnet. However, garnets in which there are rare earth ions on the c sites have a rather different magnetization curve and exhibit a compensation point where the moment due to the a and d sublattices is exactly equal to that of the c sublattice.

9.7 Interaction mechanisms

Let us now return to consider the details of the various exchange interactions.

9.7.1 Direct exchange

We have already noted that the simple Heisenberg interaction is applicable in only certain restricted crystal systems and is not widespread. It has the advantage that the relationship between the exchange integral J_e and the mean field constant can readily be determined. The Hamiltonian can be written in terms of the exchange integral as

$$H = -2J_e \sum_j \mathbf{S}_i \cdot \mathbf{S}_j = -2J_e \sum_j (S_{xi}S_{xj} + S_{yi}S_{yj} + S_{zi}S_{zj}). \quad (9.58)$$

The Ising model assumes that the terms $S_{xi}S_{xj}$ and $S_{yi}S_{yj}$ can be neglected and thus the Ising Hamiltonian is just

$$H = -2J_e \sum_j S_{zi}S_{zj}. \quad (9.59)$$

As the magnetization, assumed along the z direction, is $M = Ng_S\mu_B S$, we have for a crystal structure with p nearest neighbours

$$H = -2J_e p S_{zi} M / Ng\mu_B. \tag{9.60}$$

In terms of the mean field parameter, the Hamiltonian is

$$H = -\lambda g_S S_{zi} \mu_B \tag{9.61}$$

and hence we have

$$\lambda = 2pJ_e / Ng_S^2 \mu_B^2. \tag{9.62}$$

9.7.2 Superexchange

In antiferromagnetic and ferrimagnetic oxides we find that the antiparallel ordering of moments is not usually between nearest neighbours. Take MnO as a simple example. Here alternate Mn^{2+} ions along the cube edges are ordered antiparallel but these are separated by the large O^{2-} ions. The Mn ions are second nearest neighbours. Clearly the exchange interaction must be mediated by the oxygen ions. This effect is known as *superexchange*. The Hamiltonian has the same form as the Heisenberg Hamiltonian.

The basic principles are seen from consideration of four electrons as illustrated in Fig. 9.11. The ground state has one electron on each Mn ion in states d_1 and d_2 and two electrons in identical p orbitals on the O ion. There is a certain probability that one of the oxygen p electrons

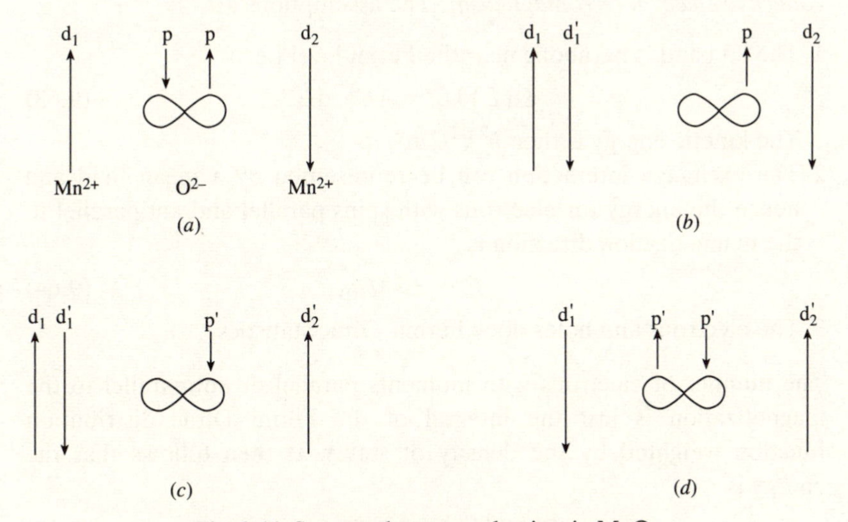

Fig. 9.11 Superexchange mechanism in MnO.

will be transferred to the Mn ion in the state d_1'. In this excited state there will be coupling between d_1 and d_1' electrons of the Mn ions and the remaining O p electron and the d_2 electron of the other Mn ion. This coupling is spin dependent. The complete exchange of electrons on the magnetic ions is accomplished by the transfer, to an excited p orbital state, of the d_1 electron. Because the p orbital dumbbell is along the line joining the next nearest neighbours, there is strong wavefunction overlap. However, the bonds joining nearest neighbour Mn ions with the neighbouring oxygen ions are at right angles. The dumbbell shaped p wavefunction of the oxygen ion only overlaps significantly with *one* of the Mn ions. Thus the next nearest neighbour interaction is strongest.

9.7.3 Exchange in transition metals

The problem here is that the electrons which are responsible for the magnetic interaction are also the conduction electrons. These electrons are *itinerant* and are not bound to a specific atom. Ferromagnetism can arise due to the fact that the energies of the bands are dependent on the electron spin. This is still a matter of extensive research, as the interaction in itinerant systems is still improperly understood and only within the last few years have people been able to put spin effects into band structure calculations with any degree of success at all. The model described below is that due to Stoner and the model is known as *collective electron ferromagnetism*. The assumptions are

1 The 3d band is parabolic near the Fermi level i.e.

$$D(E)\mathrm{d}E = AE^{1/2}\mathrm{d}E. \tag{9.63}$$

The kinetic energy is then $\hbar^2 k^2/2m^*$

2 The exchange interaction can be represented by a mean field and hence the energy for electrons with spins parallel and antiparallel to the magnetization direction is

$$E = \pm\lambda M \mu_B \tag{9.64}$$

3 The electrons and holes obey Fermi–Dirac statistics.

The number of electrons with moments parallel or antiparallel to the magnetization is just the integral of the Fermi–Dirac distribution function weighted by the density of states. It then follows that the energy is

$$E = \hbar^2 k^2/2m^* \pm \lambda M \mu_B. \tag{9.65}$$

The magnetization is μ_B times the difference in the numbers per unit volume aligned parallel and antiparallel. We have therefore

$$M = \mu_B \int \{F[E(k) - \lambda N \mu_B^2] - F[E(k) + \lambda N \mu_B^2]\} D(E) \mathrm{d}E/2 \quad (9.66)$$

where F is the Fermi–Dirac function and N is the number of 3d band electrons. This is not necessarily an integer number as the 3d and 4s bands overlap and this explains why the measured magnetic moments are non-integer values of the Bohr magneton. At absolute zero it is straightforward to evaluate the integral, namely,

$$M = [\mu_B/(2\pi)^2] \times \left[\int_0^{E_F + \lambda N \mu_B^2} (2m^*/\hbar^2) E^{1/2} \mathrm{d}E - \int_0^{E_F - \lambda N \mu_B^2} (2m^*/\hbar^2) E^{1/2} \mathrm{d}E \right]. \quad (9.67)$$

That is,

$$M = (\mu_B/6\pi^2)(2m^*/\hbar^2)^{3/2}[(E_F + \lambda N \mu_B^2)^{3/2} - (E_F - \lambda N \mu_B^2)^{3/2}]. \quad (9.68)$$

Thus we see that the magnetization is a function of the relative magnitudes of the exchange energy compared with the Fermi energy. The collective model permits non-integral values of the Bohr magneton for the effective moment per ion.

Now that we have distinct magnetic sub-bands for the 3d electrons we see quite generally how the position of the Fermi level will dictate whether a metal is ferromagnetically ordered or not. Take the example of copper, shown in Fig. 9.12(*a*). Here, the 4s and 3d bands overlap but the Fermi energy is such that the 3d bands are both full. The Fermi level electrons are thus in the 4s band and the numbers of 3d electrons parallel and antiparallel to the magnetization are equal. For nickel, which is shown schematically in Fig. 9.12(*b*), however, the Fermi level is such that one of the 3d sub-bands is full, while the other is only partly filled. Hence we have a non-zero magnetization and nickel is a ferromagnet.

9.7.4 The RKKY interaction – magnetism in the rare earths and actinides

In the rare earth metals, the conduction electrons are not those responsible for the exchange interaction. The 'magnetic' electrons are the 4f electrons which are screened from the neighbouring atoms by the 5d and 6s electrons, which also act as the conduction electrons.

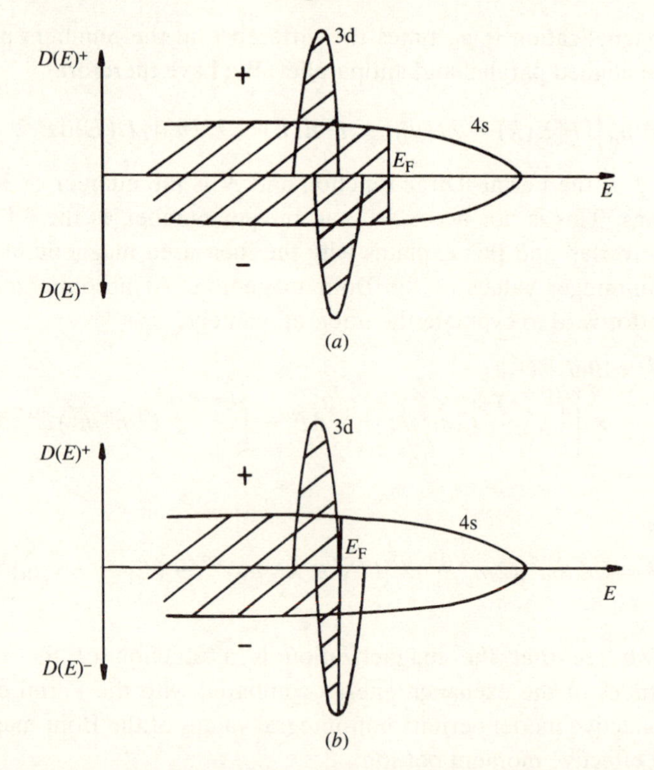

Fig. 9.12 Schematic representation of the magnetic sub-bands of (*a*) copper (*b*) nickel showing the relative position of the Fermi surface with respect to the top of the 3d bands.

There is thus very little direct overlap of the wavefunctions of the 4f electrons on adjacent atoms. The mechanism by which the exchange is mediated is indirect and involves polarization of the conduction electrons by the (almost) localized 4f electrons. The theory was worked out by Ruderman, Kittel, Kasuya and Yoshida and the interaction is thus known as the RKKY interaction.

We consider in Fig. 9.13 a single localized moment at an ion surrounded by a sea of itinerant conduction electrons. It is favourable for a conduction electron with moment parallel to the localized moment to be near to the ion. This is done by the electron wavefunction becoming distorted so that the electron density at the ion is increased. In order for this to happen, there takes place a mixing of other states of the same spin orientation. The mixing of wavefunctions results in constructive interference such that the electron concentration at the

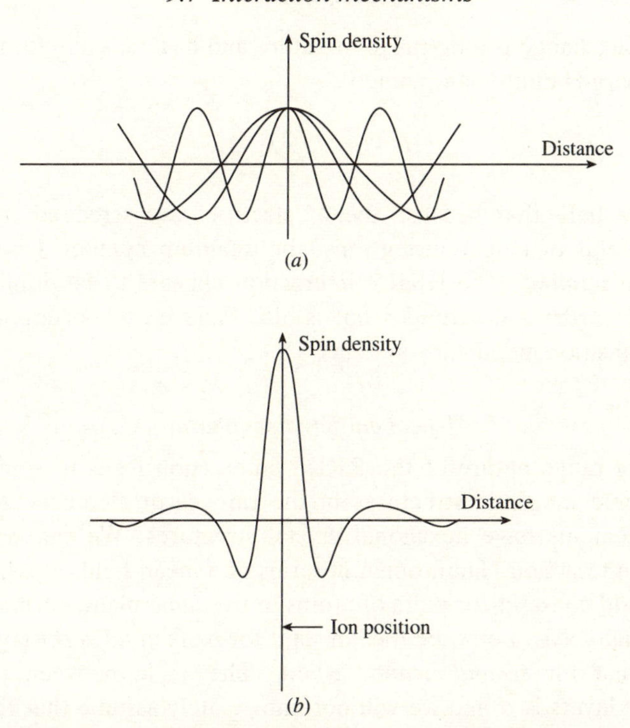

Fig. 9.13 Schematic representation of the spin density around a magnetic ion. (*a*) Wavefunctions of intermixing electron states near the ion, (*b*) resulting spin density of the conduction electrons. The oscillating nature of the spin density results in the RKKY interaction. (After J. Crangle, *The Magnetic Properties of Solids* (Edward Arnold, London: 1977).)

magnetic ion is increased. However, because the electron wavevectors differ, there will be destructive interference at some distance from the magnetic ion. Again, there will be a further distance at which constructive interference again occurs and so on with decreasing magnitude. There is thus an oscillatory electron density of parallel spin electrons. Conversely, it is unfavourable for the spin down electrons to be near the magnetic ion (with up moment) and there is a similar, and complementary, oscillation in the antiparallel electron density. As the total numbers of electrons are constant the charge distribution remains uniform. The oscillation period is determined by the electron wavevectors at the Fermi surface. Another magnetic ion at a distance r from the first, will interact either ferro- or antiferromagnetically with the first depending on whether it is located in a positive or negative part of the polarization wave from the first ion. The important point to note is that the exchange is long-range in nature and has the same form as the Heisenberg Hamiltonian, namely,

$$H = -2J_e \sum_j \mathbf{J}_i \cdot \mathbf{J}_j. \tag{9.69}$$

We note here that because the 4f electrons are screened from the crystal fields of neighbouring ions, the quantum number **J** is a good quantum number. The RKKY interaction appears to be dominant in the rare earths and actinides but is also believed to be important in some transition metal alloy systems.

Helical antiferromagnetism

The long range nature of the RKKY interaction leads to some quite remarkable magnetic structures in the rare earth elements. Most of these elements have hexagonal crystal structures. We can write the RKKY interaction Hamiltonian in terms of a mean field. Let λ_0 be the mean field constant for pairs of atoms in the *same* plane normal to the hexagonal c axis. Let λ_1 be the constant for pairs in *adjacent* layers and λ_2 be that for *second nearest* layers. The angle between spins in adjacent layers is α and we will not immediately assume that the spins in adjacent layers are parallel or antiparallel. The mean field at layer 0 will be

$$B_E = \lambda_0 M_0 + 2\lambda_1 M_1 \cos\alpha + 2\lambda_2 M_2 \cos 2\alpha \tag{9.70}$$

where M_0, M_1 and M_2 are the magnetizations of layers 0, 1 and 2. If all layers have the same magnetization, the mean field is

$$B_E = M(\lambda_0 + 2\lambda_1 \cos\alpha + 2\lambda_2 \cos 2\alpha). \tag{9.71}$$

The angle between layers will be such as to maximize the exchange interaction, i.e. maximize the mean field. Thus we have for equilibrium angle α_0,

$$\cos\alpha_0 = -\lambda_1/4\lambda_2. \tag{9.72}$$

If $|\lambda_1/4\lambda_2| < 1$ we have a stable helical state such as that shown in Fig. 9.14, where there is a constant turn angle between successive layers of aligned spins. This results in a net zero magnetization and is an antiferromagnetic structure. Suppose the period of the turn is such that in distance q there is a 2π turn, (i.e. $2\pi/\alpha_0 = q/c$, where c is the interplanar spacing). The magnetic neutron scattering factor will be modulated by the repeat distance q and we find satellite peaks appearing in the neutron diffraction pattern on either side of the magnetic

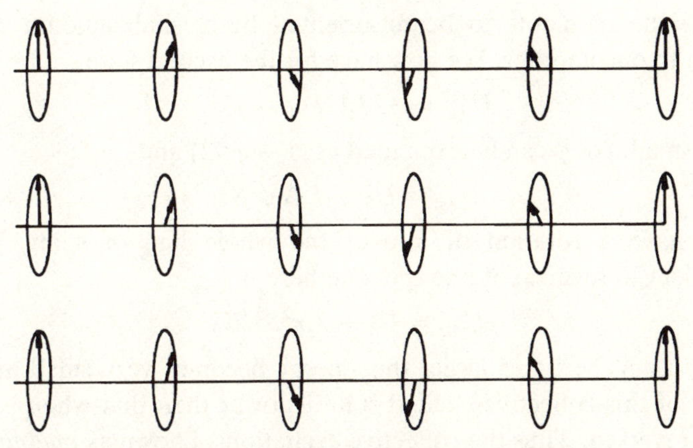

Fig. 9.14 Schematic diagram of the helical antiferromagnetic structure such as that found in dysprosium between 86 and 178 K.

Bragg peaks. For neutron wavelength λ and nuclear reflection Bragg angle θ_B they are separated by $\Delta\theta$ such that

$$\lambda = 2\Delta\theta(\cos\theta_B)q. \tag{9.73}$$

9.8 Collective magnetic excitations

We noticed earlier that the mean field approximation did not accurately predict the magnetization as a function of temperature very close to absolute zero. This is because the disordering process is assumed to be one in which single, individual spins are reversed by the thermal excitation. Let us assume a Heisenberg Hamiltonian between N spins of moment S in a line or ring. Then

$$H = -2J_e\sum_j \mathbf{S}_i \cdot \mathbf{S}_j. \tag{9.74}$$

If we treat the spins $\mathbf{S}$ as classical vectors we have for the ground state

$$U_0 = 2NJ_eS^2. \tag{9.75}$$

If one spin S_0 is reversed, the excited state has additional energy

$$U_+ = -2(N-1)J_eS^2 + 2J_eS^2 = U_0 + 4J_eS^2 \tag{9.76}$$

and the energy is increased by $4J_eS^2$.

Now if all the spins share the reversal, i.e. the excitation is a collective one, then the energy of the first excited state is lower than that given by Equations (9.75) and (9.76). The collective excitation permits the moments to be misoriented by a small amount ϕ with respect to one another. We now have for the excited state

$$U_{+c} = -2J_e(N\cos\phi)S^2. \tag{9.77}$$

If ϕ is small, $\cos\phi$ can be expanded as $(1 - \phi^2/2)$ and

$$U_{+c} = U_0 + J_e N\phi^2 S^2. \tag{9.78}$$

If we have a rotation of π over the whole line of spins, this is equivalent to reversal of one spin and here

$$U_{+c} = U_0 + J_e\pi^2 S^2/N. \tag{9.79}$$

Clearly as N becomes large, this energy becomes very small and the energy of this collective excited state is lower than that when a single spin is reversed. Thus the collective excitations, known as *magnons* are responsible for the disordering of the moments as the temperature rises from $T = 0$.

9.8.1 Magnon dispersion relation

In just the same way as one may consider the collective excitation of a lattice in the form of a phonon, we can consider the collective excitation of the spin system as a magnon. The classical derivation of the magnon dispersion relation is found by calculating the torque on spin p due to adjacent spins and equating that to the rate of change of angular momentum, as in the derivation of the Bloch equations (Equations (8.12)). For a line of atoms in the z direction, for the spin components S_p^x, S_p^y and S_p^z of atom p, one obtains a set of simultaneous equations of the form

$$\mathrm{d}S_p^x/\mathrm{d}t = (2JS/\hbar)(2S_p^y - S_{p-1}^y - S_{p+1}^y), \tag{9.80a}$$

$$\mathrm{d}S_p^y/\mathrm{d}t = (2JS/\hbar)(2S_p^x - S_{p-1}^x - S_{p+1}^x), \tag{9.80b}$$

$$\mathrm{d}S_p^z/\mathrm{d}t = 0. \tag{9.80c}$$

We look for solutions in the form of a travelling wave

$$S_p^x = u\exp[\mathrm{i}(pka - \omega t)] \quad \text{and} \quad S_p^y = v\exp[\mathrm{i}(pka - \omega t)]$$

where u and v are constants and p is an integer. We have two simultaneous equations and in order that there should be a non-trivial solution, we require that the determinant of the coefficients must vanish. Hence we find that

$$\hbar\omega = 4JS(1 - \cos ka) \tag{9.81}$$

is the dispersion relation. For small k this approximates to

$$\hbar\omega = (2JSa^2)k^2. \tag{9.82}$$

In the long wavelength limit, it is straightforward to show that in a three-dimensional cubic lattice with nearest neighbour interaction, the frequency is again proportional to k^2. (Note that for phonons, in the long wavelength limit, the frequency is proportional to k, not k^2.)

The displacement of the moments with respect to each other is shown in Fig. 9.15. The magnon dispersion relation can be determined by inelastic neutron scattering, just as the phonon dispersion relation can be determined. Note that spin waves (magnons) are stiffer than phonons and so we need more energetic neutrons to excite the magnons.

9.8.2 Low temperature magnetization

We can calculate the low temperature magnetization in the same manner that we calculate the phonon density for thermal transport properties. In thermal equilibrium, the average number of magnons excited in the mode of wavevector **k** is given by the Bose–Einstein distribution, namely

$$n(\mathbf{k}) = [\exp(\hbar\omega/k_B T) - 1]^{-1}. \tag{9.83}$$

The total number of magnons excited at temperature T is then

$$n = \int D(\omega)n(\omega)\mathrm{d}\omega \tag{9.84}$$

where $D(\omega)$ is the number of modes in frequency range between ω and $\omega + \mathrm{d}\omega$.

Proceeding as for the density of states of electrons in a three-dimensional metal, we have that the number of modes between wavevector k

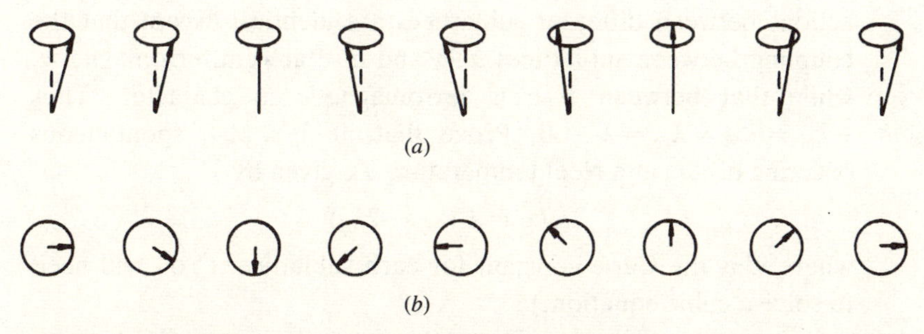

Fig. 9.15 Relative displacements of magnetic moments in a magnon or spin wave. (*a*) Viewed from the side, (*b*) viewed from above.

and $k + \mathrm{d}k$ is $(1/2\pi)^3 4\pi k^2 \mathrm{d}k$. Thus the number of modes between ω and $\omega + \mathrm{d}\omega$ is

$$\mathrm{D}(\omega)\mathrm{d}\omega = (1/2\pi)^3(4\pi k^2)(\mathrm{d}\omega/\mathrm{d}k)\mathrm{d}\omega. \tag{9.85}$$

Hence as $\mathrm{d}\omega/\mathrm{d}k = 4JSa^2k/\hbar^2$ we have

$$\mathrm{D}(\omega) = (1/4\pi^2)(\hbar/2JSa^2)^{3/2}\omega^{1/2}. \tag{9.86}$$

Thus, the total number of magnons n is

$$n = (1/4\pi^2)(\hbar/2JSa^2)^{3/2}\int_0^{\omega_{\max}} [\exp(\hbar\omega/k_\mathrm{B}T) - 1]^{-1}\omega^{1/2}\mathrm{d}\omega. \tag{9.87}$$

That is, writing x as a dummy variable

$$n = (1/4\pi^2)(k_\mathrm{B}T/2JSa^2)^{3/2}\int_0^{\infty} [\mathrm{e}^x - 1]^{-1}x^{1/2}\mathrm{d}x. \tag{9.88}$$

In the low temperature limit the upper limit of the integral may be set to infinity and hence the number of magnons excited varies as $T^{3/2}$. The difference in magnetization ΔM from that at absolute zero $M(0)$ is thus proportional to $T^{3/2}$ as is found experimentally.

It is extremely interesting to see that a calculation, essentially the same as that performed to calculate the density of states of free electrons in an ideal metal can be used to explain the properties of totally different types of material in an apparently unrelated branch of physics.

Problems

9.1 A material has atoms with identical spins on three different lattice sites, a, b and c. The mean field constants for interactions between atoms of the same sublattice are equal and ferromagnetic in character, i.e. $\lambda_{aa} = \lambda_{bb} = \lambda_{cc} = \lambda_{ii} > 0$. The constants for interactions between different sublattices are identical except that the coupling between sublattices a–b and a–c are antiferromagnetic, while that between b–c is ferromagnetic in character. Thus $-\lambda_{bc} = \lambda_{ab} = \lambda_{ac} = \lambda_{ij} < 0$. Prove that, if $\lambda_{ii} < |\lambda_{ij}|$, spontaneous ordering occurs at a Néel temperature T_N given by

$$T_\mathrm{N} = C(\lambda_{ii} - 2\lambda_{ij})$$

where C is the Curie constant for each sublattice. (You will need to solve a cubic equation.)

9.2 Spin waves can be excited in antiferromagnets in a similar manner to ferromagnetic magnons. Let spins of even indices $2p$ compose sublattice $a(S_z = +S)$ and those of odd indices $2p + 1$ compose sublattice $b(S_z = -S)$. Consider only nearest neighbour interactions, let the exchange integral J_e be negative and assume that the spins never move far from alignment in the $\pm z$ direction.

(a) Prove that the classical torque equations for the a spins are

$$\mathrm{d}S^x_{2p}/\mathrm{d}t = (2J_eS/\hbar)(-2S^y_{2p} - S^y_{2p-1} - S^y_{2p+1})$$
$$\mathrm{d}S^y_{2p}/\mathrm{d}t = -(2J_eS/\hbar)(-2S^x_{2p} - S^x_{2p-1} - S^x_{2p+1})$$

and for the b spins are

$$\mathrm{d}S^x_{2p+1}/\mathrm{d}t = (2J_eS/\hbar)(+2S^y_{2p+1} + S^y_{2p} + S^y_{2p+2})$$
$$\mathrm{d}S^y_{2p+1}/\mathrm{d}t = -(2J_eS/\hbar)(+2S^x_{2p+1} + S^x_{2p} + S^x_{2p+2}).$$

(b) Show that by combining in the form $S^+ = S^x + \mathrm{i}S^y$ and looking for solutions of the type

$$S^+_{2p} = u \exp(\mathrm{i}pka - \mathrm{i}\omega t) \qquad S^+_{2p+1} = v \exp(\mathrm{i}pka - \mathrm{i}\omega t)$$

one arrives at a dispersion relation of the form

$$\omega = \omega_{ex}|\sin ka|.$$

(c) Derive an expression for the density of antiferromagnons $D(\omega)\mathrm{d}\omega$ between ω and $\omega + \mathrm{d}\omega$.

(d) Use this result to prove that the magnetic contribution to the specific heat capacity of an antiferromagnet varies as T^3 at very low temperatures.

9.3 The dispersion relation for low wavevector magnons on a ferromagnetic lattice is

$$\omega = Dk^2$$

where ω is the angular frequency, k the wavevector, and D the stiffness constant. Prove that the magnetic contribution to the specific heat capacity of a ferromagnet varies as $T^{3/2}$ at a very low temperatures.

9.4 A ferrimagnet has sublattice magnetizations M_a, M_b and M_c associated with three different lattice sites. In the mean field approximation, the constants describing the coupling between spins on the a and b sites, the a and a sites, the b and b sites and the c and c sites are λ_{ab}, λ_{aa}, λ_{bb} and λ_{cc} respectively. The ac and bc constants $\lambda_{ac} = \lambda_{bc} = 0$. Using the same approach as for the antiferromagnet, derive an expression for the ferrimagnetic Néel temperature in terms of the mean field constants λ_{ij} (where i and j can stand for a, b or c). Comment on the form of the result.

9.5 (a) Prove that, in the mean field approximation, the contribution to the specific heat capacity associated with ferromagnetic ordering is

$$C_M = [3JNk_B/2(J+1)]\mathrm{d}\{[M(T)/M(0)]^2\}/\mathrm{d}(T/T_c)$$

where $M(T)$ and $M(0)$ are the magnetizations at temperatures T and 0 respectively.

(b) Using the form of $M(T)$ as a function of T, sketch the form of this curve and comment on its shape.

(c) Prove that the height of the discontinuity at T_c is given by

$$C_M = 5J(J+1)Nk_B/[J^2 + (J+1)^2].$$

(d) Prove that at very low temperatures

$$C_M = [4M(0)^2 T_c/T^2]\exp(-2T_c/T)[1 - 2\exp(-2T_c/T)].$$

10
Superconductivity

Probably the most spectacular phenomenon associated with the breakdown of the independent electron approximation is that of superconductivity. In the superconducting state, the material loses all resistivity and becomes a perfect conductor. The discovery in 1986 of oxide materials which were superconducting at temperatures above that of the boiling point of nitrogen sparked an unprecedented surge of activity in the field which remains an area of high profile and popular interest.

10.1 The discovery of superconductivity

In 1908 Kammerlingh Onnes succeeded in liquefying helium and set about the task of studying the properties of metals at these extremely low temperatures. As we saw in Chapter 1, the resistivity of metals such as platinum fell to a small, non-zero value when extrapolated to $T = 0$. This residual resistivity fell with improvements in purity and thus Onnes studied mercury, which was the most pure metal available at that time. To the great surprise of Onnes and the whole scientific community, the resistivity fell monotonically until just above the boiling point of helium and then fell abruptly to zero. Fig. 10.1 shows an example of the superconducting phase transition in yttrium barium copper oxide, one of the high temperature superconducting oxides. Onnes was unable to measure precisely the transition width or whether the resistivity was genuinely zero. However, in 1963 File and Mills measured the decay of a persistent current set up in a superconducting ring using nuclear magnetic resonance as the probe. They found a lower limit for the decay time of 100 000 years, indicating a ratio of less than 10^{-15} between the resistivities in the superconducting and normal

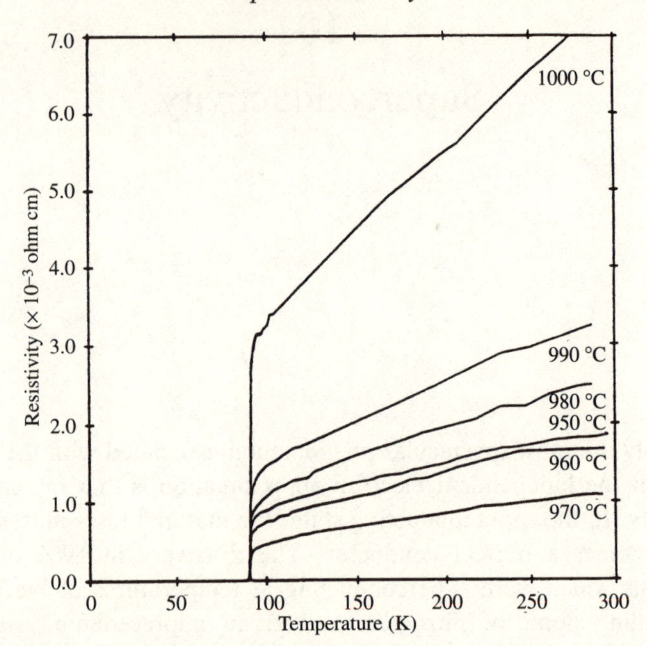

Fig. 10.1 Superconducting transition in yttrium barium copper oxide ($YBa_2Cu_3O_{7-\delta}$). Resistance as a function of temperature for polycrystalline pellets sintered at different temperatures. (From M. R. Delap, Ph.D. Thesis, University of Durham, 1990.)

states. Subsequently 27 elements and an effectively unlimited number of inorganic and organic compounds have been found to be superconducting. It is surprising to note that the materials do not have to be extremely pure for the superconducting transition to take place. Lead gas pipes are found to be just as good superconductors as highly purified specimens of lead. The inorganic compounds listed in Table 10.1 are those with the highest transition temperatures with the exception of the oxide superconductors (Table 10.2).

10.2 Magnetic properties

10.2.1 Type 1 superconductors

It was found that the elemental superconductors behaved in a very different way from some of the compound superconductors when placed in an external magnetic field. When an elemental superconductor is placed in a magnetic field, it will only remain superconducting up to a certain value of field, before abruptly reverting to the normal state

Table 10.1 *Superconducting transition temperatures of some 'low T_c' compounds*

Nb_3Ga	21.0 K	V_3Ga	16.8 K
Nb_3Sn	18.05 K	Ta_3Pb	16.0 K
Nb_3Ge	23.2 K	NbN	17.2 K
Mo_3Re	15.0 K	MoC	14.3 K

Table 10.2 *Superconducting transition temperatures of some 'high T_c' oxides*

$YBa_2Cu_3O_{7-\delta}$	90 K	$Bi_2Sr_2Ca_2Cu_3O_{10}$	110 K
$La_{1.85}Sr_{0.15}CuO_4$	37 K	$Tl_2Ba_2Ca_2Cu_3O_{10}$	122 K
$Ba_{0.6}K_{0.4}BiO_3$	30 K	$Tl_2Ba_2CaCu_2O_8$	110 K
$HgBa_2Ca_2Cu_3O_{8+\delta}$	133.5 K	$(Pb, Hg)Sr_2(Ca, Y)Cu_2O_7$	90 K

at a critical field B_{0c}. This critical field is a function of temperature and varies as

$$B_{0c} = B_{0c}(0)[1 - (T/T_c)^2] \tag{10.1}$$

where $B_{0c}(0)$ is the critical field extrapolated to absolute zero and T_c is the transition temperature in zero external field. This is a general equation valid for all the superconducting elements.

The Meissner effect

Perhaps the most unexpected phenomenon associated with superconductivity was the Meissner effect. In the normal state the type 1 superconductors have only a small susceptibility and external flux passes through the material, effectively without change in the flux density (Fig. 10.2(*a*)). However, it was found that when a superconductor was cooled below its transition temperature in a very small applied field, the flux through the materials was abruptly excluded (Fig. 10.2(*b*)). The material becomes a perfect diamagnet. Screening currents flow in the surface of the superconductor to shield the interior and we will discuss these in detail in a later section. The flux density in the material **B**, the flux density in free space $\mathbf{B}_0$, the magnetizing force **H** and the magnetization M are related by

$$\mathbf{B} = \mu_0(\mathbf{H} + \mathbf{M}) = \mathbf{B}_0 + \mu_0\mathbf{M} \tag{10.2}$$

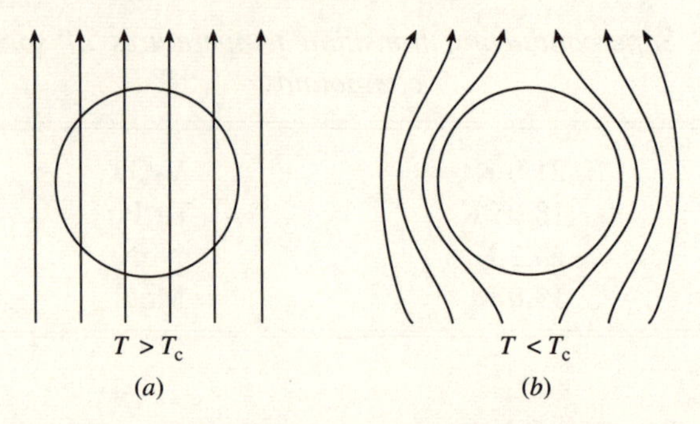

Fig. 10.2 The Meissner effect. (*a*) Flux passing through material in the normal state. (*b*) Flux totally expelled on passing into the superconducting state.

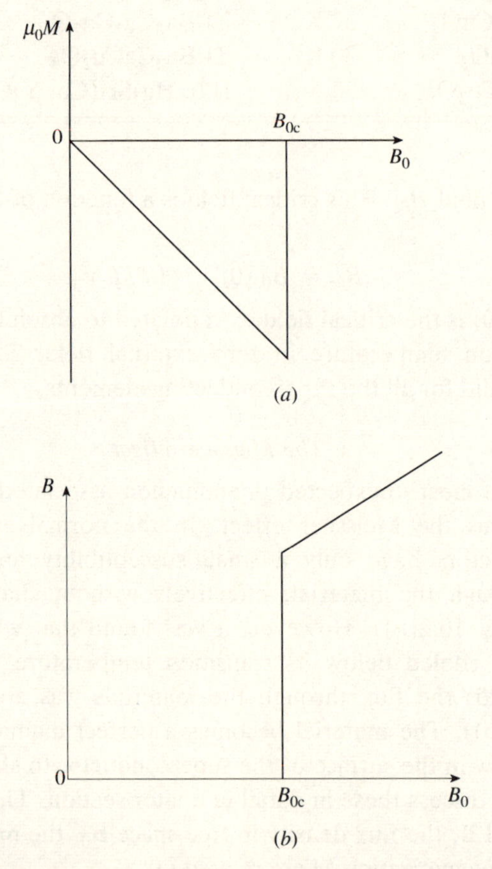

Fig. 10.3 (*a*) Magnetization and (*b*) internal flux denisty as a function of applied field in a type 1 superconductor.

where μ_0 is the permeability of free space. We see that in the superconducting state the type 1 superconductor has

$$\mathbf{M} = -\mathbf{B}_0/\mu_0 \tag{10.3}$$

and $\mu_0 k = -1$. (k is the susceptibility). This is illustrated in Fig. 10.3 where the effect of the critical field B_{0c} is also shown. Note that the flux inside the specimen is zero for applied fields less than B_{0c}. (It is worth noting that the measured value of B_{0c} does depend on specimen shape, because the demagnetizing field is as important in the superconducting state as in the ferromagnetic state. This must be included when calculating the internal flux density.) Levitation of a superconductor can occur in a magnetic field due to the high diamagnetic susceptibility. If the superconductor is displaced, the lines of flux excluded from the material are compressed. This increases the energy of the external field and thus a restoring force is exerted on the body.

It is important to stress that the phenomenon of flux exclusion is *not* a necessary result of the existence of zero resistance, but a totally separate feature of the superconducting state.

10.2.2 Type 2 superconductors

The critical field of type 1 superconductors is found to be small and thus, because of the magnetic field associated with passage of an electric current, the current density which may be carried in the material before it reverts to the normal state is small. It was the discovery of superconducting compounds which had very different magnetic properties which led the way to the development of superconducting magnets where extremely high supercurrents flow without loss of energy and hence without heating the magnet. These type 2 superconductors, as they were called, exhibit a complete Meissner effect at low fields and are perfectly diamagnetic (Fig. 10.4(*a*)). However, at the critical field, $B_{0c}(1)$ the flux does not completely enter the specimen. Instead the flux enters in flux lines of normal material contained within the superconducting matrix (Fig. 10.4(*b*)). As we will see later, these flux lines consist of a single quantum of magnetic flux Φ_0, given by

$$\Phi_0 = h/2e. \tag{10.4}$$

Because of the mutual repulsion of the flux lines, they form a two-dimensional hexagonal array, shown in Fig. 10.4(*c*). The flux lines can be revealed by evaporating fine iron particles onto the surface of the superconductor. The particles are attracted to regions of field maxima and delineate the flux line lattice, making the flux line positions visible in a transmission electron microscope.

This mixed state, also known as the 'vortex state', may persist over a very large field range. As the external field is increased so the flux line density increases so that more flux penetrates the sample. At a second critical field $B_{0c}(2)$ flux fully penetrates the sample and the material reverts to the normal state. For materials such as Nb–Sn alloys, this upper critical field is extremely high and superconducting magnets based on this material have been made which give steady fields of 20 tesla. On passage of an electrical current of density J, the flux lines experience a force of $\mathbf{J} \times \mathbf{\Phi}_0$, which tends to make them move. This flux motion leads to an induced voltage, which in turn leads to energy dissipation through an effective resistance. Flux line motion is detected by the presence of hysteresis in the magnetization curves for type 2

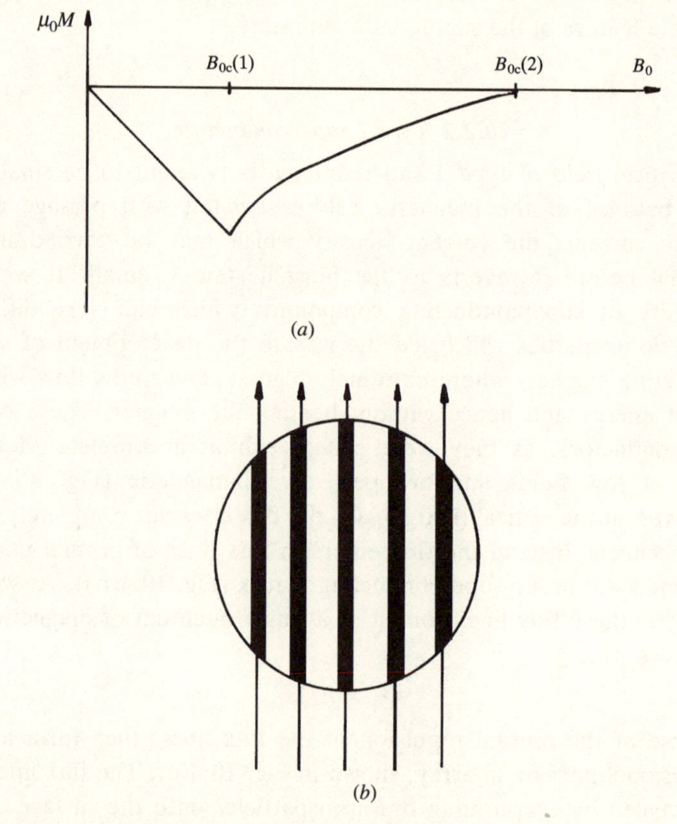

Fig. 10.4 (*a*) and (*b*). See facing page for caption.

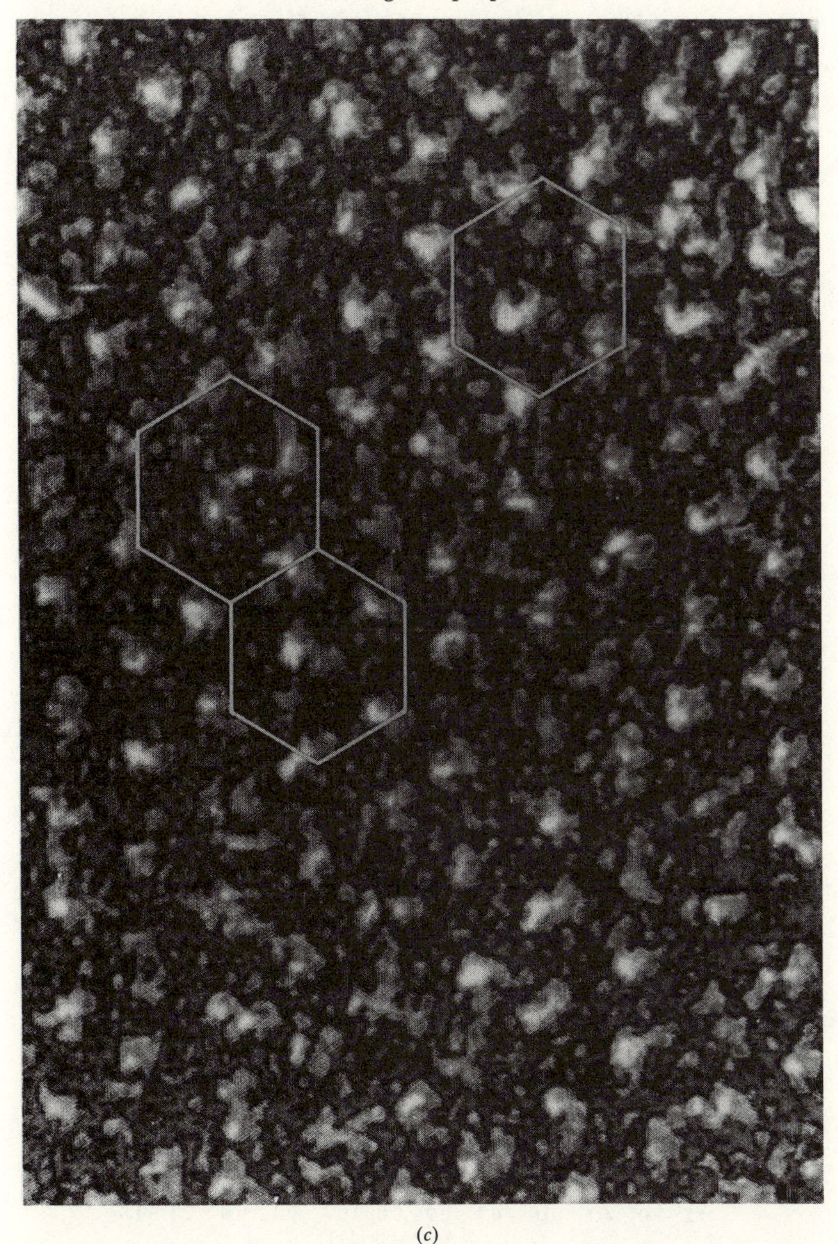

(c)

Fig. 10.4 (*a*) Magnetization as a function of applied field for a type 2 superconductor. (*b*) The vortex state where flux penetrates the superconductor in a hexagonal array of flux lines. (*c*) Electron microscope image of ferromagnetic iron particles decorating the flux line lattice in $Bi_2Sr_2CaCuO_x$. The flux lattice is viewed normal to the applied field. (After P. L. Gammel, C. A. Murray and D. J. Bishop, *Mater. Res. Bull.* **15** (1990) 31.)

superconductors (Fig. 10.5). It is essential, for the manufacture of technologically useful materials, to pin these flux lines. This is achieved by the creation of point defects in the alloying process, control of grain size and work hardening to obtain flux line pinning at the grain boundaries and dislocations.

10.3 Thermal properties

The thermal properties of superconductors can be explained phenomenologically using thermodynamic treatments. Consider a superconductor of volume V in a magnetic field B_0 which is increased such that the magnetization is gradually increased. The infinitesimal amount of work $\mathrm{d}W$ done on increasing the magnetization from M to $M + \mathrm{d}M$ is

$$\mathrm{d}W = -VB_0\mathrm{d}M. \tag{10.5}$$

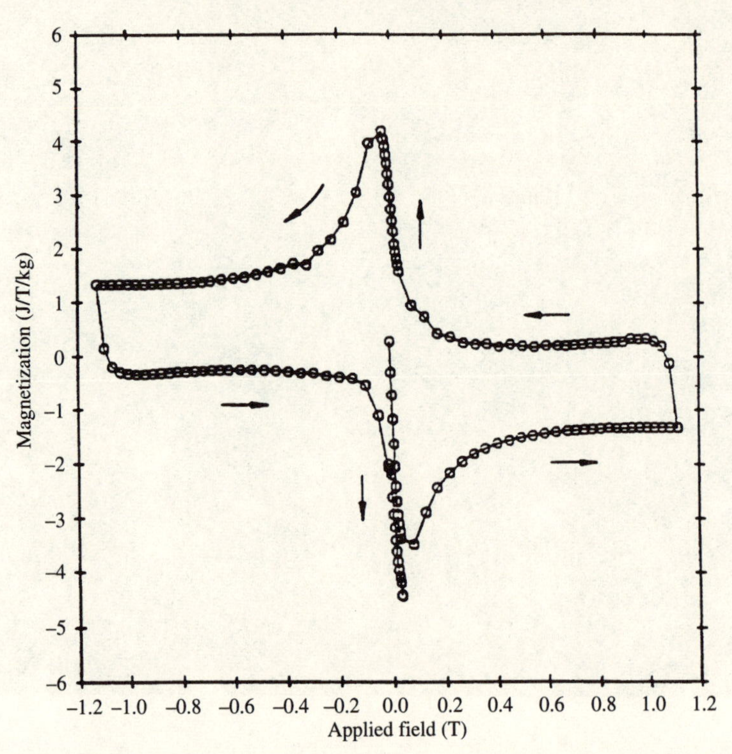

Fig. 10.5 Magnetic hysteresis curve taken at 40 K for the type 2 superconductor yttrium barium copper oxide. The area under the hysteresis loop corresponds to the work done per unit volume on taking the material around a full magnetization cycle. (After M. R. Delap, Ph.D. Thesis, University of Durham, 1990.)

The Gibbs free energy G is defined as

$$G = U - TS - VB_0M \tag{10.6}$$

where U is the internal energy and S is the entropy. Now from the First Law of thermodynamics,

$$\mathrm{d}U - T\mathrm{d}S - VB_0\mathrm{d}M = 0 \tag{10.7}$$

so

$$\mathrm{d}G = -S\mathrm{d}T - VM\mathrm{d}B_0. \tag{10.8}$$

For a perfect diamagnet

$$\mu_0 M = -B_0 \tag{10.9}$$

and we therefore have

$$G_\mathrm{s}(B_0) = G_\mathrm{s}(0) + VB_0^2/2\mu_0 \tag{10.10}$$

where $G_\mathrm{s}(B_0)$ and $G_\mathrm{s}(0)$ are the Gibbs free energies in the superconducting state, at fields of B_0 and zero respectively. Now in the normal state a superconductor is essentially non-magnetic and

$$G_\mathrm{n}(B_0) = G_\mathrm{n}(0) = G_\mathrm{n}. \tag{10.11}$$

At the critical field $B_{0\mathrm{c}}$, the Gibbs free energy is equal in the normal and superconducting states and thus

$$G_\mathrm{n}(B_{0\mathrm{c}}) = G_\mathrm{s}(B_{0\mathrm{c}}). \tag{10.12}$$

Therefore

$$G_\mathrm{n} = G_\mathrm{s}(0) + VB_{0\mathrm{c}}^2/2\mu_0. \tag{10.13}$$

The critical field is therefore determined by the difference in the free energies of the normal and superconducting states.

Let us now suppose that the phase transition takes place at $B_{0\mathrm{c}} - \mathrm{d}B_{0\mathrm{c}}$ and $T_\mathrm{c} - \mathrm{d}T_\mathrm{c}$. Then

$$G_\mathrm{s} - \mathrm{d}G_\mathrm{s} = G_\mathrm{n} - \mathrm{d}G_\mathrm{n}. \tag{10.14}$$

Thus

$$-S_\mathrm{s}\mathrm{d}T - VM_\mathrm{s}\mathrm{d}B_{0\mathrm{c}} = -S_\mathrm{n}\mathrm{d}T - VM_\mathrm{n}\mathrm{d}B_{0\mathrm{c}}. \tag{10.15}$$

As $M_\mathrm{n} = 0$ and $M_\mathrm{s} = -B_{0\mathrm{c}}/\mu_0$, we have

$$S_\mathrm{s} - S_\mathrm{n} = VB_{0\mathrm{c}}(\mathrm{d}B_{0\mathrm{c}}/\mathrm{d}T)/\mu_0. \tag{10.16}$$

The latent heat of transition L is therefore

$$L = -VTB_{0\mathrm{c}}(\mathrm{d}B_{0\mathrm{c}}/\mathrm{d}T)/\mu_0. \tag{10.17}$$

At $T = T_\mathrm{c}$, $B_{0\mathrm{c}} = 0$ and thus in zero field, there is no latent heat associated with the superconducting phase transition. This type of transition is known as a second order phase transition. The specific heat capacity C is given by

$$C = T(\mathrm{d}S/\mathrm{d}T) \tag{10.18}$$

and so the difference in heat capacity of the superconducting and normal states is

$$\begin{aligned} C_s - C_n &= T[(\mathrm{d}S_s/\mathrm{d}T) - (\mathrm{d}S_n/\mathrm{d}T)] \\ &= VT[B_{0c}(\mathrm{d}^2 B_{0c}/\mathrm{d}T^2) + (\mathrm{d}B_{0c}/\mathrm{d}T)^2]/\mu_0. \end{aligned} \tag{10.19}$$

At $T = T_c$, $B_{0c} = 0$ and

$$C_s - C_n = \{[VT(\mathrm{d}B_{0c}/\mathrm{d}T)^2]/\mu_0\}_{T=T_c}. \tag{10.20}$$

There is thus a discontinuity in the specific heat capacity at the superconducting phase transition (Fig. 10.6(*a*)) in zero field. The magnitude of this anomaly is found to be in good agreement with Equation (10.20) for the metallic superconductors. At low temperatures, the specific heat capacity of a normal metal varies as $C = AT + BT^3$ where A and B are constants, whereas in the superconducting state (Fig. 10.6(*b*)), the electronic heat capacity varies as $\exp(-2\Delta/k_B T)$. This is suggestive of the existence of an energy gap of magnitude Δ.

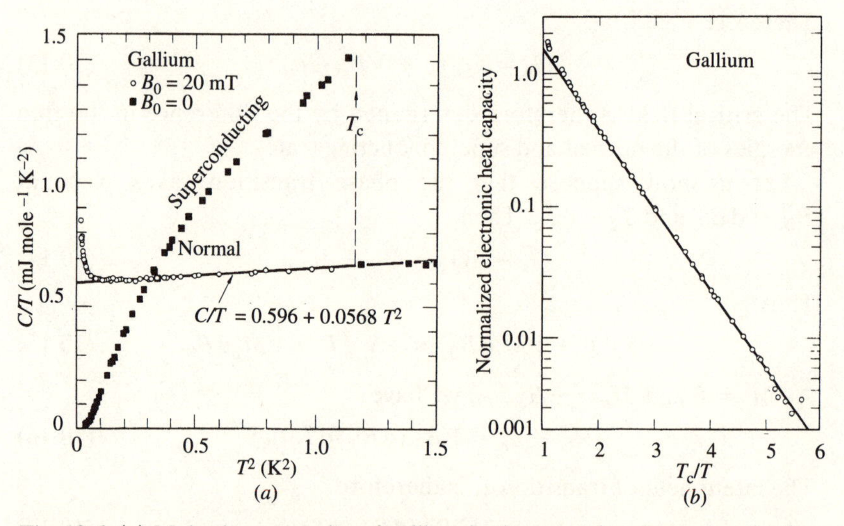

Fig. 10.6 (*a*) Molar heat capacity of gallium in the normal and superconducting states. (*b*) Electronic part of the heat capacity in the superconducting state showing the exponential dependence on inverse temperature. (After N. E. Philips, *Phys. Rev.* **134** (1964) 385.)

10.4 Microwave and infra-red absorption

The suspicion that an energy gap is present in the superconducting state is reinforced by the observation of the absorption of microwaves as a function of temperature. Microwaves can be transmitted through thin foils of metal, provided that the thickness is less than about 2 nm. If the film is cooled through the superconducting transition temperature, the transmission rises, indicating that now the microwave photons do not have enough energy to excite electrons across an energy gap. Similarly, if the infra-red reflectivity is measured as a function of frequency from a material in the superconducting state, a sharp increase in reflectivity occurs at a specific frequency ω, again suggestive of an energy gap given by

$$\hbar\omega = 2\Delta. \tag{10.21}$$

10.5 Single particle tunnelling

If a superconductor and a normal metal are separated by a thin insulating layer, provided that the insulating layer is comparable with the conduction electron wavelength then electrons can quantum-mechanically tunnel through the barrier. (An example of such a system is a bead of the common lead–tin solder on a constantan wire. The constantan oxidizes readily and forms a thin insulating barrier. When the solder is cooled to liquid helium temperature, it becomes superconducting and a tunnelling current can be observed under an external voltage bias.) If the flow is metal–insulator–superconductor, Ohm's Law is obeyed, but if the flow is in the opposite sense, at very low temperatures no current flows until a critical voltage is reached given by $eV = \Delta$ (Fig. 10.7). At higher temperature, current does flow below this value, but there is still a discontinuity at the critical voltage.

10.6 Theoretical models of superconductivity

10.6.1 Microscopic theory

From the above discussion it is clear that, somehow, the electron scattering processes present in a normal metal, which lead to the resistivity, are being suppressed. It is also clear that an energy gap exists, although this cannot be tied to the crystal lattice periodicity in the manner of the band gaps discussed earlier. The key piece of

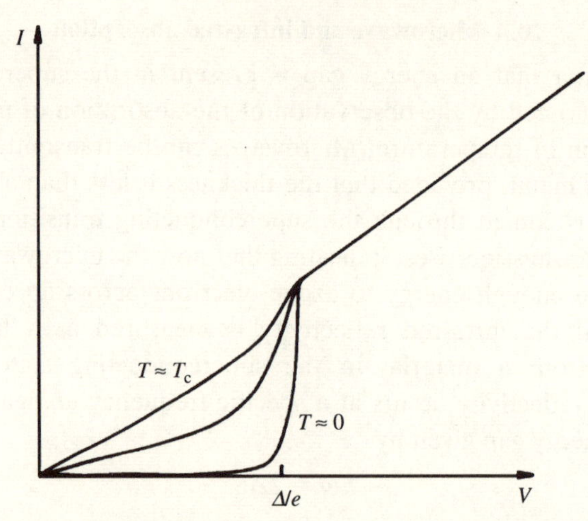

Fig. 10.7 Current–voltage plots for a superconductor–insulator–normal metal system at different values of temperature.

experimental evidence came from a study of the value of the superconducting phase transition for different isotopes of the same element. It was found that, for conventional low temperature superconductors such as mercury, lead and niobium,

$$M^{-1/2}T_c = \text{constant} \tag{10.22}$$

where M is the isotopic mass. In analogy to a mass on a spring, the vibrational frequency of the crystal lattice varies as $M^{-1/2}$, and this suggests that the quantized lattice vibrations, or 'phonons' are in some way responsible for the occurrence of superconductivity. Phonon involvement is further supported by the observation that the elemental superconductors with high transition temperatures have relatively high room temperature resistivities.

In 1950, Frohlich suggested that the motion of an electron in a crystal could perturb the lattice in such a way as to cause attraction of another electron. This process can be thought of as emission and absorption of a virtual phonon in much the same way that electromagnetic forces can be considered as the emission and absorption of virtual photons. An alternative picture is one of the deformation of the lattice by the Coulomb attraction due to the electron charge. As illustrated in Fig. 10.8(a), there is a net repulsive force between conduction electrons. This is less than a bare Coulomb force due to the screening of the positive ionic charge. Assuming that the lattice is not completely rigid, each electron attracts the positive ions and deforms the lattice,

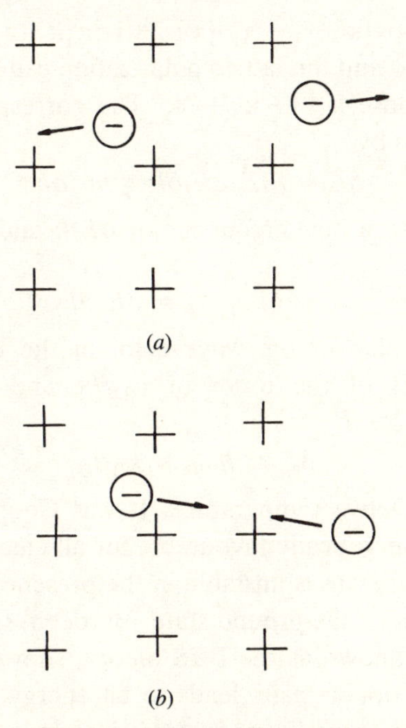

Fig. 10.8 (*a*) Repulsive interaction between electrons in an undeformed lattice. (*b*) Deformation and polarization of the ions to yield a net attractive interaction.

polarizing it (Fig. 10.8(*b*)). The extent of this depends on the lattice stiffness. Under some circumstances, the deformation may be such that the net charge seen by an electron in the proximity of another is positive, and thus the interaction is attractive.

In general this interaction is not attractive as the lattice polarization, equivalent to a phonon, moves with velocity of the order of the velocity of sound v_s. Conduction electrons move with velocity v_F, the velocity at the Fermi surface. However, if we consider a pair of electrons with opposite wavevectors $\mathbf{k}_1$ and $\mathbf{k}_2$, the wavefunction of the pair can be written

$$\psi_{k_1k_2(r_1,r_2)} = 2^{-1/2}[\exp(\mathrm{i}\mathbf{k}_1\cdot\mathbf{r}_1)\exp(\mathrm{i}\mathbf{k}_2\cdot\mathbf{r}_2) \pm \exp(\mathrm{i}\mathbf{k}_1\cdot\mathbf{r}_2)\exp(\mathrm{i}\mathbf{k}_2\cdot\mathbf{r}_1)]. \tag{10.23}$$

Still, in general the pair will travel with velocity near v_F unless $\mathbf{k}_1 = -\mathbf{k}_2 = \mathbf{k}$. Then the pair forms a standing wave as

$$\psi_{k,-k(r_1,r_2)} = 2^{-1/2}\{\exp[\mathrm{i}\mathbf{k}\cdot(\mathbf{r}_1-\mathbf{r}_2)] \pm \exp[-\mathrm{i}\mathbf{k}\cdot(\mathbf{r}_1-\mathbf{r}_2)]\}. \tag{10.24}$$

This is either $2^{1/2} \cos [\mathbf{k} \cdot (\mathbf{r}_1 - \mathbf{r}_2)]$ or $2^{1/2} \mathrm{i} \sin [\mathbf{k} \cdot (\mathbf{r}_1 - \mathbf{r}_2)]$. The group velocity is thus zero and the lattice polarization can be induced.

Suppose we choose $\mathbf{k}_1 = -\mathbf{k}_2 + \delta\mathbf{k}$. The corresponding energy difference δE is given by

$$\delta E = (\partial E/\partial k)\delta k = \hbar v_\mathrm{F} \delta k. \tag{10.25}$$

This corresponds to a beat frequency of $\delta E/\hbar$ and a beat velocity v_B given by

$$v_\mathrm{B} = \delta k v_\mathrm{F}/k_\mathrm{F} = \delta E/\hbar k_\mathrm{F}, \tag{10.26}$$

where k_F is the value of the wavevector at the Fermi surface. The phonon velocity is of the order of $\omega_\mathrm{D}/k_\mathrm{F}$ and thus an attractive interaction is possible if

$$\delta E \ll \hbar\omega_\mathrm{D} \equiv k_\mathrm{B}\theta_\mathrm{D} \tag{10.27}$$

where θ_D is the Debye temperature. It was Cooper who showed in 1950 that it was energetically favourable for all electrons to pair in this matter. The normal state is unstable in the presence of this interaction and the paired state is the ground state. Bardeen, Cooper and Schrieffer, in the theory known as the BCS theory, showed that the binding energy of these 'Cooper' pairs leads to an energy gap 2Δ opening in the density of states at the Fermi surface given by

$$\Delta = \hbar\omega_\mathrm{D} \exp\{-[g(E_\mathrm{F})\mathcal{G}]^{-1}\} = 7k_\mathrm{B}T_\mathrm{c} \tag{10.28}$$

where $g(E_\mathrm{F})$ is the density of states at the Fermi surface and $\mathcal{G}$ is a measure of the electron–phonon interaction. The exponential term is large and T_c is thus much less than θ_D. Equation (10.28) shows why the isotope effect occurs, as ω_D is proportional to $M^{-1/2}$. When an electric field is applied, each Cooper pair has non-zero momentum and a current flow takes place. Due to the presence of the energy gap at the Fermi surface, this cannot be stabilized by normal scattering processes and the current flows without resistance. An alternative simple model for the existence of zero resistance is illustrated in Fig. 10.9. Consider a free electron E–k distribution (Fig. 10.9(*a*)). When an electric field is applied, the distribution shifts and scattering provides stability by the transfer of electrons from states such as *a* to *b* (Fig. 10.9(*b*)). However, in the superconducting state, we can consider the binding of the Cooper pairs as 'springs' as in Fig. 10.9(*c*). Now when the whole distribution is displaced on setting up the drift velocity, transfer from state *a* to state *b* can only occur if the bond between the Cooper pairs is broken. If the current density and temperature are sufficiently low this does not occur (Fig. 10.9(*d*)) and the current flows with no loss.

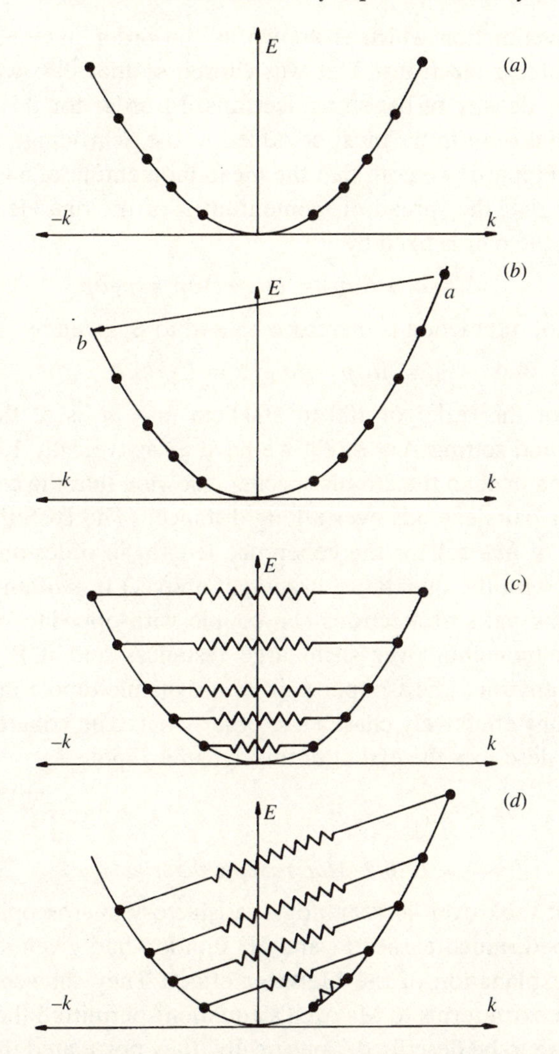

Fig. 10.9 Schematic model of the coupling of Cooper pairs. (*a*) Free electron distribution. (*b*) Non-equilibrium distribution under current flow. (*c*) Binding of Cooper pairs of opposite wavevector. (*d*) Inability of scattering processes to equilibrate the distribution.

An important property of the Cooper pairs is the distance over which the interaction effectively takes place. Pippard has given a direct physical interpretation of this coherence length based on the uncertainty principle but it also emerges from the independent theory of Ginsburg and Landau. (Their approach was to introduce a pseudo-wavefunction which is known as the order parameter for the superconducting electrons. This was chosen so that $|\Psi^2|$ was equal to the number density of the superelectrons. In order for this to happen an additional term in Ψ^3 must be added to the Schrödinger equation.) Following Pippard, we note that the mean momentum of a Cooper pair is zero but that the spread of momentum p of the one electron levels making up the pair is fixed by

$$\Delta = \delta(p^2/2m) = (p_F/m)\delta p = v_F \delta p. \tag{10.29}$$

The range of interaction is therefore limited to ξ_0 given by

$$\xi_0 = \hbar/\delta p = \hbar v_F/\Delta = E_F/k_F \Delta. \tag{10.30}$$

As E_F is of the order of 100 to 1000 nm and k_F is of the order of 10^{-10} m^{-1} and setting $\Delta \approx k_B T_c$, we have ξ_0 as typically 100 nm. This is much greater than the atomic spacing, showing that the coherence of the Cooper pairs extends over a long distance. (The BCS theory leads to a value of $\hbar v_F/\pi\Delta$ for the coherence length, in order-of-magnitude agreement with the qualitative argument above.) It is often difficult to imagine how pairs of electrons can couple with opposite wavevectors and hence momenta over such large distances and it is sometimes helpful to envisage the Cooper pair as a dynamic dipole in which the two electrons effectively chase each other's tail. The coherence length effectively describes the extent of this dynamic dipole.

10.6.2 Macroscopic theories

Although it took over 40 years for a satisfactory microscopic theory to be developed, much earlier F. and H. London had given a phenomenological explanation of the Meissner effect. They showed that addition of two extra terms to Maxwell's equations permitted the superconducting state to be described. Specifically, they postulated that

$$\partial \mathbf{J}/\partial t = \mathbf{E}/\mu_0\lambda^2 \quad \text{and} \quad \nabla \times \mathbf{J} = -\mathbf{B}/\mu_0\lambda^2, \tag{10.31}$$

where λ is a characteristic length.

Then as

$$\nabla \times \mathbf{H} = \mathbf{J} \tag{10.32}$$

we have

$$\nabla \times \mathbf{J} = \nabla \times \nabla \times \mathbf{H} = -\nabla^2\mathbf{H} = -\nabla^2\mathbf{B}/\mu_0. \tag{10.33}$$

Therefore

$$\nabla^2\mathbf{B} = \mathbf{B}/\lambda^2. \tag{10.34}$$

Let us now consider a semi-infinite type 1 superconducting solid extending for $x > 0$. The interface with a vacuum is at $x = 0$ and contains the (yz) plane. Let the field outside the superconductor be in the z direction and magnitude B_0. Solution of Equation (10.33) then reduces to a one-dimensional problem and we have for the region $x > 0$ (inside the superconductor)

$$B(x) = A \exp(x/\lambda) + C \exp(-x/\lambda), \tag{10.35}$$

where A and C are constants. In order that the solution shall not diverge as x goes to infinity, the coefficient A must be zero. Thus the flux density inside the superconductor decays as

$$B(x) = B_0 \exp(-x/\lambda). \tag{10.36}$$

Clearly λ is the characteristic decay length inside the superconductor and it is known as the London penetration depth. Thus the expulsion of flux from the body of the superconductor occurs over a small distance comparable with λ. Within this surface region, screening currents flow to compensate exactly the external flux density and prevent penetration of flux into the bulk of the superconductor.

10.7 Type 1 and type 2 superconductors

The twin concepts of coherence length and penetration depth allow us to explain the difference in behaviour of the two classes of superconductor. In order for the vortex state to exist in the type 2 superconductor, there must be many interfaces between the normal and superconducting state. These interfaces are not microscopically abrupt; the magnetic flux penetrates the superconducting region over a distance of the order of λ and the superconducting state persists into the normal region over a distance of the order of ξ_0. If the total energy associated with the exclusion of field and the formation of the superconducting state is positive, then interfaces are not favoured, and the material is a type 1 superconductor. Conversely, if the presence of an interface results in a net reduction in the free energy, then the material exhibits type 2 behaviour.

Consider an element of interface, assumed planar. As seen from Equation (10.13), the increase in energy due to the destruction of superconductivity in the coherence length ξ_0 is $B_{0c}^2\xi_0/2\mu_0$ per unit area (Fig. 10.10). The loss in field energy due to exclusion of the field in depth λ is $B_{0c}^2\lambda/2\mu_0$ per unit area. Thus the total interface energy is $B_{0c}^2(\xi_0 - \lambda)/2\mu_0$ per unit area. Thus if $\xi_0 > \lambda$ we have type 1 behaviour

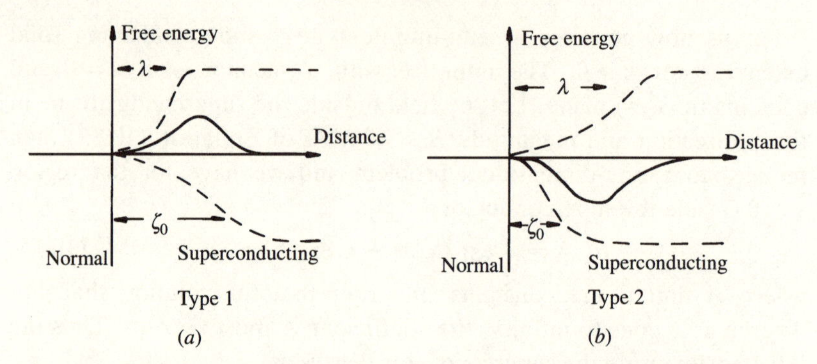

Fig. 10.10 Free energy associated with a planar interface between the superconducting and normal states. (*a*) Type 1 superconductor (*b*) type 2 superconductor.

and if $\xi_0 < \lambda$ we have type 2 behaviour. The type 2 material should enter the vortex state at a critical field B_{01} given by

$$(B_{0c}^2\xi_0/2\mu_0) - (B_{01}^2\lambda/2\mu_0) \approx 0. \tag{10.37}$$

Thus the lower critical field B_{01} where the Meissner effect breaks down and the vortex state is entered is given by

$$B_{01} = (\xi_0/\lambda)^{1/2} B_{0c}. \tag{10.38}$$

For the high temperature superconductors such as yttrium barium copper oxide, $\xi_0 \ll \lambda$ and thus the lower critical field B_{01} is extremely small (typically mtesla) while the upper critical field B_{0c} is many tens of tesla.

10.8 Flux quantization

A very simple analysis shows that a superconducting ring has quite remarkable properties in relation to the flux which threads through it. Consider an open loop of normal conductor, cross-sectional area A, through which a flux threads. If the flux changes, a voltage V is induced across the ends of the loop given by Faraday's Law

$$V = -\partial\Phi/\partial t. \tag{10.39}$$

If the end of the loop were closed, and the resistance of the loop were R, then a current I would flow given by

$$I = -(\partial\Phi/\partial t)/R. \tag{10.40}$$

If the loop is superconducting, *either* the current becomes infinite for a small non-zero change in the flux through the loop *or*, if the material remains superconducting, the flux through the loop cannot change. The former is impossible and indeed we find that flux is conserved through a superconducting loop.

The long range quantum nature of the superconducting state leads also to the concept of flux quantization. In the Ginzburg–Landau formalism, a pseudo-wavefunction Ψ is associated with each Cooper pair. This is chosen such that $\Psi^*\Psi = n_s$, the density of Cooper pairs. Let us consider a superconducting material in which n_s is constant. Then, as at $T = 0$ n_s is half the number of electrons in the conduction band, we may write

$$\Psi = n_s^{1/2} \exp[i\theta(r)] \quad \text{and} \quad \Psi^* = n_s^{1/2} \exp[-i\theta(r)]. \tag{10.41}$$

In the general case we can write, for particle velocity

$$\mathbf{v} = (\mathbf{p} - q\mathbf{A})/m \tag{10.42}$$

where m is the particle mass, q the charge, $\mathbf{A}$ the magnetic vector potential and $\mathbf{p}$ the momentum. The particle flux is then

$$\Psi^*\mathbf{v}\Psi = n_s(\hbar\nabla\theta - q\mathbf{A})m \tag{10.43}$$

where we have replaced the momentum $\mathbf{p}$ by its operator equivalent, $-i\hbar\nabla$ and we have dropped the r dependent symbol from θ. An immediate consequence of the fact that the coupling of the individual electrons in $(\mathbf{k}, -\mathbf{k})$ pairs is that the net pair flux is zero and thus

$$\hbar\nabla\theta = q\mathbf{A}. \tag{10.44}$$

Now let us consider a loop of superconducting wire through which threads a flux Φ. We may now integrate the pair phase along elements of path $\mathbf{dl}$ through the body of the superconducting loop giving

$$\int \theta \cdot \mathbf{dl} = \theta_2 - \theta_1 \tag{10.45}$$

for the phase change on going once around the ring and returning to the identical start position. As the pseudo-wavefunction must be single valued, we have that

$$\theta_2 - \theta_1 = 2\pi s \tag{10.46}$$

where s is an integer. Stoke's theorem enables us to integrate the right hand side of Equation (10.44) over the area enclosed by the superconducting loop. Explicitly,

$$\int \mathbf{A} \cdot \mathbf{dl} = \int (\nabla \times \mathbf{A}) \cdot \mathbf{dS} = \int \mathbf{B} \cdot \mathbf{dS} = \Phi. \tag{10.47}$$

Thus we find that the flux threading a superconducting loop must be

$$\Phi = \hbar 2\pi s/q = hs/2e \tag{10.48}$$

as the charge of the Cooper pair is $2e$. Equation (10.48) states that the flux through a superconducting ring must be *quantized* in units of the flux quantum $\Phi_0 = h/2e = 2.0678 \times 10^{-15}$ tesla m^2. As the flux lines penetrating a type 2 superconductor are essentially normal regions surrounded by a superconducting loop, it now becomes clear why the flux lines in the vortex state contain one flux quantum.

10.9 The SQUID magnetometer

We have already considered the remarkable properties associated with tunnelling of electrons through a thin insulating barrier between a superconductor and a normal metal. Even more remarkable effects are observed when a thin insulating barrier separates two superconducting materials. Then, provided that the kinetic energy of the carriers is not high, i.e. at low current densities, the Cooper pairs can cross the insulating barrier from one superconductor to the other without loss of phase coherence in the pseudo-wavefunction Ψ. This leads to a number of remarkable effects, for example that a non-zero d.c. current flows across such a junction in the presence of zero applied potential difference (the d.c. Josephson effect) and that an a.c. current flows in the presence of a d.c. potential difference across the junction (the a.c. Josephson effect). These all arise from the fact that there is a well defined phase difference of the pair on either side of the junction, the tunnelling current being related to the magnitude of this phase difference. These effects have led to the construction of a number of devices which exploit the Josephson effect to measure magnetic flux very precisely (SQUIDs or superconducting quantum interference devices). The most common of these is the r.f. SQUID.

Let us consider a loop of superconductor in which is what is known as a 'weak link'. This is a very thin insulating barrier, which might be formed, for example, by evaporation of an insulating oxide or a point contact such as that illustrated in Fig. 10.11(a). At low currents, the Cooper pairs will tunnel through the insulating barrier and the loop will behave as a superconductor. However, if the current in the loop exceeds a critical current, then the weak link ceases to preserve the phase coherence and the loop reverts to the normal state. Let us suppose that the loop, containing the weak link, encloses a flux Φ. This must be an integral number, s, of flux quanta Φ_0. Let us now increase the external flux to an arbitrary value Φ_{ext}. In order that the flux enclosed by the ring shall remain at $s\Phi_0$, a screening current I_s

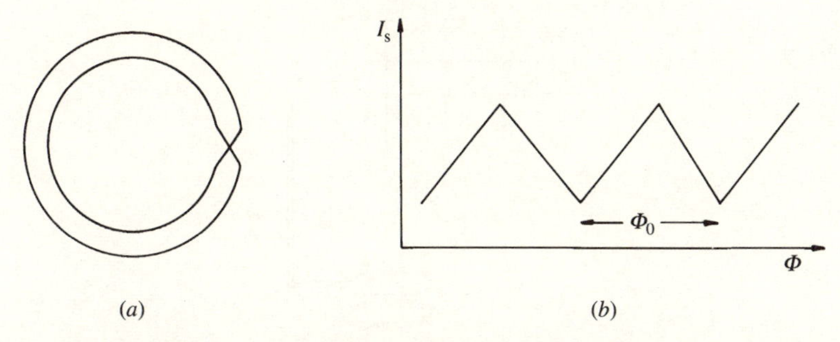

Fig. 10.11 The point contact SQUID. (*a*) Schematic construction. (*b*) Screening current as a function of external magnetic flux.

flows in the surface (i.e. within the London penetration depth) of the superconductor given by

$$\Phi_{ext} + LI_s = s\Phi_0 \qquad (10.49)$$

where L is the inductance of the loop. (The weak link will in general also result in a Josephson current flowing and this adds an additional term to the left hand side of Equation (10.49). It does not however, affect our subsequent discussion.)

As the external flux is increased, so the screening currents increase until the weak link goes normal at the critical current. Flux can then penetrate the ring. However, as soon as one flux quantum enters the ring, the screening current drops below the critical value and the loop again becomes superconducting. The screening current is thus a periodic function of the external flux, as shown schematically in Fig. 10.11(*b*). For a small change in external flux, the screening current changes linearly and we thus have a means of measuring extremely small values of magnetic flux. It is usual to couple flux into a SQUID by means of a superconducting flux transformer. This exploits yet again the fact that the flux through a superconducting loop cannot be changed provided that the loop remains superconducting. If we have an arrangement such as that shown in Fig. 10.12(*a*), a change in the flux at the large diameter external loop will result in a change in the screening current. This results in a change in flux at the SQUID. Clearly by use of a large area external loop, very small flux densities can be measured, such as those associated with the operation of the human brain. Use of two external coils in opposition gives sensitivity only to field gradients (Fig. 10.12(*b*)). In order to sense the screening currents, it is usual to incorporate the single junction SQUID into a radio frequency circuit. SQUIDs are finding increasing application in

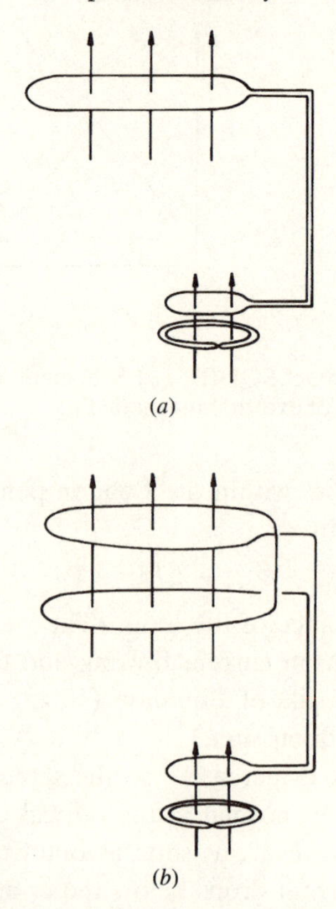

Fig. 10.12 Superconducting flux transformer to measure (*a*) magnetic field (*b*) magnetic field gradient.

detection of small magnetic fields in such widely disparate areas as archaeology, medicine, and non-destructive testing as well as basic physics. The recent development and commercial availability of SQUIDs based on the high T_c oxides which work at liquid nitrogen temperature will certainly extend greatly their range of application.

Problems

10.1 Using Equations (10.1) and (10.17) prove that at temperature $T_c/2$, the latent heat absorbed when superconductivity is destroyed by application of a magnetic field is $3B_0(0)^2/8\mu_0$.

10.2 A superconducting material has critical temperature T_c, normal state specific heat capacity $C_n = \gamma T$ per unit volume and critical field given by Equation (10.1). Prove that the entropy density of a thin superconducting rod in a field $B_0 < B_{0c}$ applied parallel to the rod is

$$S_s = \gamma T - 2B_{0c}^2(0)(T/T_c^2)[1 - (T/T_c)^2]/\mu_0.$$

What is the entropy density for $B_0 > B_{0c}$?

Show that adiabatic magnetization of the superconductor (i.e. slow increase of B_0 from zero to above $B_{0c}(0)$ on a thermally isolated sample) provides a mechanism for low temperature refrigeration. Prove that the maximum temperature drop occurs for an initial temperature of $T_c/3^{1/2}$. (You may assume $\mu_0\gamma > 2B_{0c}^2(0)/T_c^2$.)

10.3 A superconducting magnet consists of 15000 turns of wire wound into a solenoid of diameter 3 cm and length 0.1 m. Calculate the energy dissipated if the wire abruptly 'quenches' from the superconducting to the normal state from a field of 18 tesla. Compare this result with that of Problem 10.1. If the latent heat of liquid helium is 2.5 J cm^{-3}, calculate the volume of liquid boiled off in a quench. (It is alarming!!)

10.4 Using Maxwell's equations and the London equations (Equations (10.31)) prove that an electric field outside a type 1 superconductor decays into the sample with a characteristic decay length λ.

10.5 A thin plate of thickness d and very large superficial area is placed in a field B_0 and cooled below the superconducting phase transition. Prove that, if the field is parallel to the plate, the average magnetization in the plate M is given by

$$M = [(2\lambda/d)\tanh(d/2\lambda) - 1]B_0/\mu_0.$$

What would happen if the field was perpendicular to the plate?

10.6 A SQUID magnetometer is used to study the change in magnetization of a spherical sample in a zero external field between temperatures of 2 and 6 K. At 2 K it is a single domain ferromagnetic particle with saturation magnetization 5×10^5 A m^{-1}. At 5 K it undergoes a transition to the paramagnetic state. The sample is located at the centre of a single turn of niobium wire of diameter 1 cm which is coupled to the SQUID as a superconducting flux transformer. If it is required that the SQUID should be able to read the change in magnetization continuously without resetting, determine the largest diameter particle that can be used.

If the minimum possible sample diameter which you could use

in practice was 3×10^{-6} m, would you increase or decrease the niobium loop diameter in order to achieve a continuous reading?

10.7 A certain superconductor has a penetration depth of 2.5×10^{-7} m and a Fermi energy of 10 eV. At very low temperatures, infra-red radiation is very well transmitted down to a wavelength of 130 μm below which the transmission drops sharply. Determine whether this material might be worth developing for production of superconducting solenoids.

Appendix 1
Elements of kinetic theory

We consider a cubic box, sides x, y, z, of classical particles each of mass m moving randomly. Let the velocity of particle number i be v_i and the x, y and z components of that velocity be v_{ix}, v_{iy} and v_{iz}. Now when the particle collides, perfectly elastically, with the wall of the container forming the (yz) plane, the change of momentum will be $2mv_{ix}$.

Now suppose there are N_{ix} particles per unit volume with velocity component v_{ix}. Then in a cylinder length $v_{ix}\delta t$ and of unit cross-sectional area, there are $N_{ix}v_{ix}\delta t$ particles. In time δt half of these will collide with the wall (see Fig. A1.1), so the number of particles striking the wall in unit time is $N_{ix}v_{ix}/2$.

Pressure is force per unit area, and from Newton's Second Law, force equals rate of change of momentum. Therefore, we have, for the pressure p_{ix} on the wall (yz) due to the particles of velocity component v_{ix}

$$p_{ix} = 2mv_{ix}N_{ix}v_{ix}/2 = mN_{ix}v_{ix}^2. \tag{A1.1}$$

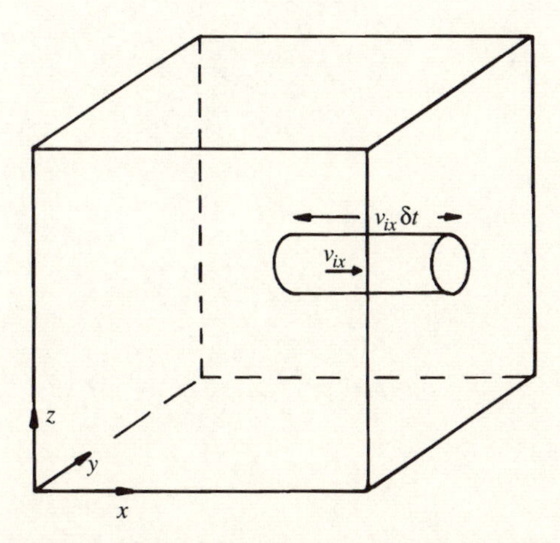

Fig. A1.1 Schematic representation of the cylinder of length $v_{ix}\delta t$ containing the particles striking the wall in time δt.

The average of the component of the velocity squared, the mean square velocity in the x direction is

$$\langle v_{ix}^2 \rangle = \sum_i N_{ix} v_{ix}^2 / N \tag{A1.2}$$

where N is the total number of particles per unit volume.

Now we could just as well have followed the argument for the y or z components of velocity and thus

$$\langle v_{ix}^2 \rangle = \langle v_{iy}^2 \rangle = \langle v_{iz}^2 \rangle. \tag{A1.3}$$

As the pressure will be the same on all walls we have

$$\langle v_{ix}^2 \rangle + \langle v_{iy}^2 \rangle + \langle v_{iz}^2 \rangle = \langle v^2 \rangle \tag{A1.4}$$

$$\langle v_{ix}^2 \rangle = \langle v^2 \rangle / 3 \tag{A1.5}$$

where $\langle v^2 \rangle$ is the mean square velocity.

Thus the pressure p is given by

$$p = Nm \langle v^2 \rangle / 3. \tag{A1.6}$$

The root mean square velocity V is defined as

$$V = (\langle v^2 \rangle)^{1/2}. \tag{A1.7}$$

Using the perfect gas law equation (A1.6) yields

$$mV^2/3 = k_B T. \tag{A1.8}$$

Appendix 2

Elements of statistical mechanics

The basic postulate of statistical mechanics is that all ways of arranging particles in a system, (called the complexions W), are equi-probable.

Boltzmann distribution for a solid

Assume N *distinguishable particles*. Suppose n_1 particles have energy E_1, n_2 have energy $E_2, \ldots, n_i$ have energy E_i etc. The number of complexions is the number of ways we can place N distinguishable particles into boxes labelled E_1, E_2 etc. so there are n_1, n_2, etc. in each box.

$$W = N!(n_1!n_2! \ldots n_i! \ldots) \tag{A2.1}$$

subject to

$$N = \sum_i n_i \tag{A2.2}$$

and the total energy E given by

$$E = \sum_i n_i E_i. \tag{A2.3}$$

The most probable state is that where W is a maximum. For convenience only we look for the maximum of $\ln W$, not W. A small change in $\ln W$, $\delta(\ln W)$ is given by

$$\delta(\ln W) = \sum_i [\partial(\ln W)/\partial n_i]\delta n_i \tag{A2.4}$$

subject to the condition that N and E are constant. This latter condition requires

$$\sum_i \delta n_i = 0, \tag{A2.5}$$

$$\sum_i E_i \delta n_i = 0. \tag{A2.6}$$

The method of undetermined multipliers may be readily found in mathematical texts and this gives

$$\delta(\ln W) + \alpha\sum_i \delta n_i + \beta\sum_i E_i \delta n_i = 0 \tag{A2.7}$$

where α and β are undetermined multipliers independent of n_i. Hence

$$\sum_i [\partial(\ln W)/\partial n_i + \alpha + E_i\beta] = 0. \tag{A2.8}$$

This must be true for all n_i, however, and thus

$$\partial(\ln W)/\partial n_i + \alpha + E_i\beta = 0. \tag{A2.9}$$

Now Stirling's theorem gives

$$\ln N! = (N + \tfrac{1}{2})(\ln N) - N + [\ln(2\pi)]/2. \tag{A2.10}$$

Therefore we have

$$\ln N! = N \ln N - N \tag{A2.11}$$

if N is very large. Then

$$\ln W = \ln N! - \sum_i [n_i \ln(n_i) - n_i]. \tag{A2.12}$$

Therefore

$$\partial(\ln W)/\partial n_i = -\partial(n_i \ln n_i - n_i)/\partial n_i = -\ln n_i \tag{A2.13}$$

and

$$-\ln n_i + \alpha + \beta E_i = 0, \tag{A2.14}$$

which gives

$$n_i = \exp(\alpha)\exp(\beta E_i) = A \exp(\beta E_i), \tag{A2.15}$$

where A is a constant. This is the classic Boltzmann distribution. If we use the Boltzmann equation relating entropy S to disorder, namely

$$S = k_B \ln W \tag{A2.16}$$

it is straightforward to show that the constant β is given by

$$\beta = -1/k_B T. \tag{A2.17}$$

The Fermi–Dirac distribution

We saw in Chapter 2 that the consequence of the quantum mechanical boundary conditions is that there are only certain allowed energy states which can be occupied by the electrons in a metal. We also noted that for a macroscopic material, the spacing of the energy levels is very small and forms a quasi-continuum. For convenience, we group the energy levels together into bundles such that there are g_k levels in the kth bundle. Let there be n_k particles in these g_k levels. Now the Pauli exclusion principle requires that there is only *one particle per level* (excluding spin degeneracy). The number of arrangements is then

$$g_k(g_k - 1)(g_k - 2)\ldots(g_k - n_k + 1) = g_k!/(g_k - n_k)!.$$

However, the particles are indistinguishable and thus the number of arrangements within the kth level is $g_k!/(g_k - n_k)!n_k!$. Therefore

$$W = g_k!/(g_k - n_k)!n_k!. \tag{A2.18}$$

Proceeding as before, using equation (A2.9), we have

$$\partial\{\ln[g_k!/(g_k - n_k)!n_k!]\}\partial n_k + \alpha + \beta E_k = 0. \tag{A2.19}$$

Using Stirling's theorem (Equation (A2.11)) we have

$$\ln(g_k - n_k)! = (g_k - n_k)\ln(g_k - n_k) - (g_k - n_k) \tag{A2.20a}$$

$$\ln n_k! = n_k \ln n_k - n_k \tag{A2.20b}$$

and thus

$$\partial[\ln(g_k - n_k)!]/\partial n_k = -\ln(g_k - n_k) \tag{A2.21a}$$

$$\partial(\ln n_k!)/\partial n_k = \ln n_k. \tag{A2.21b}$$

Therefore

$$\ln(g_k - n_k) - \ln n_k + \alpha + \beta E_k = 0, \tag{A2.22}$$

i.e.

$$(g_k/n_k) - 1 = \exp(-\alpha - \beta E_k) \tag{A2.23}$$

or, using Equation (A2.17),

$$n_k = g_k/[1 + \exp(-\alpha + E_k/k_B T)]. \tag{A2.24}$$

Setting $\alpha = E_F/k_B T$, where E_F is the Fermi energy, we have the Fermi–Dirac distribution

$$n_k = g_k/\{1 + \exp[(E_k - E_F)/k_B T]\}. \tag{A2.25}$$

This defines the probability that an indistinguishable particle subject to the Pauli exclusion principle will have energy E_k.

Appendix 3

Derivation of the Landé *g* factor

Under conditions where the orbital and spin angular momentum couple such that

$$\mathbf{J} = \mathbf{L} + \mathbf{S} \tag{A3.1}$$

the associated magnetic moments couple such that

$$\boldsymbol{\mu} = \boldsymbol{\mu}_{\mathrm{L}} + \boldsymbol{\mu}_{\mathrm{S}} \tag{A3.2}$$

However, as seen in Fig. A3.1, the moment $\boldsymbol{\mu}$ is not colinear with **J**, as the *g* factor for orbital and spin moments differs by a factor of 2. (For the orbital term $g_{\mathrm{L}} = 1$ while for the spin term $g_{\mathrm{S}} = 2$.) There is a static component $\boldsymbol{\mu}_{\mathrm{J}}$

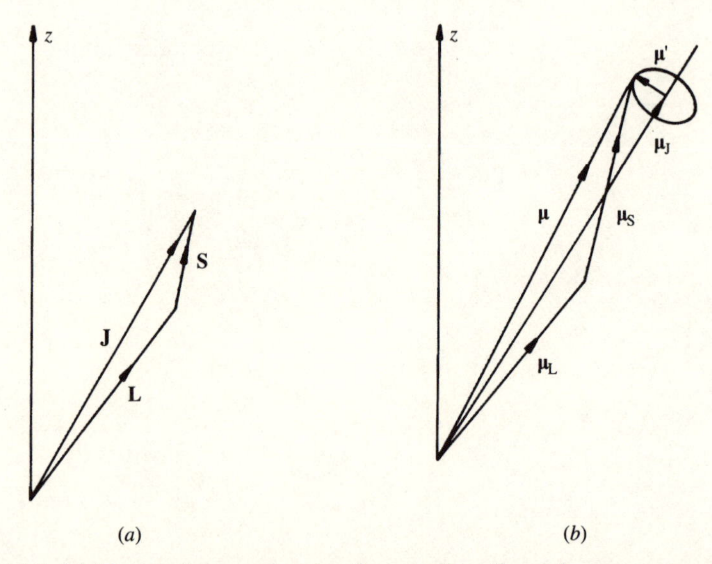

Fig. A3.1 (*a*) Vector diagram showing the coupling of angular momenta under the **L–S** coupling regime. (*b*) Corresponding diagram for the associated magnetic moments.

parallel to **J** and a component $\boldsymbol{\mu}'$ which precesses around **J**. Projecting $\boldsymbol{\mu}$ on to **J** gives, for $\boldsymbol{\mu}_J$

$$\mu_J^2 = (\mu_B/\hbar)^2(\mathbf{L}\cdot\mathbf{J}/|\mathbf{J}| + 2\mathbf{S}\cdot\mathbf{J}/|\mathbf{J}|)^2. \qquad \text{(A3.3)}$$

As

$$\mathbf{S}^2 = (\mathbf{J}-\mathbf{L})^2 = \mathbf{J}^2 + \mathbf{L}^2 - 2\mathbf{L}\cdot\mathbf{J} \qquad \text{(A3.4a)}$$

and

$$\mathbf{L}^2 = (\mathbf{J}-\mathbf{S})^2 = \mathbf{J}^2 + \mathbf{S}^2 - 2\mathbf{S}\cdot\mathbf{J}, \qquad \text{(A3.4b)}$$

we have

$$\boldsymbol{\mu}_J^2 = \left(\frac{\mu_B}{\hbar}\right)^2\left(\frac{3\mathbf{J}^2 + \mathbf{S}^2 - \mathbf{L}^2}{2|\mathbf{J}|^2}\right)^2\mathbf{J}^2. \qquad \text{(A3.5)}$$

Thus

$$\boldsymbol{\mu}_J^2 = \mu_B^2\left[\frac{3J(J+1) + S(S+1) - L(L+1)}{2J(J+1)}\right]^2 J(J+1). \qquad \text{(A3.6)}$$

Then

$$\mu_J = g_J/\mu_B[J(J+1)]^{1/2} \qquad \text{(A3.7)}$$

where

$$g_J = 1 + \frac{J(J+1) + S(S+1) - L(L+1)}{2J(J+1)}. \qquad \text{(A3.8)}$$

Index